Ralph Franklin Walworth was born in 1921 and was educated at Shrivenham College in England and the University of Southern California. He presently lives in St Petersburg, Florida.

Ralph Franklin Walworth
with
Geoffrey Walworth Sjostrom

Subdue the Earth

PANTHER
GRANADA PUBLISHING
London Toronto Sydney New York

Published by Granada Publishing Limited
in Panther Books 1980

ISBN 0 586 04682 8

A Panther UK Original

Granada Publishing Limited
Frogmore, St Albans, Herts AL2 2NF
and
3 Upper James Street, London W1R 4BP
866 United Nations Plaza, New York, NY 10017, USA
117 York Street, Sydney, NSW 2000, Australia
100 Skyway Avenue, Rexdale, Ontario, M9W 3A6, Canada
PO Box 84165, Greenside, 2034 Johannesburg, South Africa
CML Centre, Queen & Wyndham, Auckland 1, New Zealand

Set, printed and bound in Great Britain by
Cox & Wyman Ltd, Reading
Set in Intertype Times

Acknowledgements

The family and friends of authors are entitled to apologies and thanks because they accepted the behavior of the writer as the story was being prepared. The burden placed on those close to an inventor is much greater. They must bear the musings, frustrations and, worst of all, the self-centered joy that accompany the long, slow development of an idea.

During the evolution of the synthesis which is the subject of this book, those close to me were subjected to my pronouncements and radical conclusions for nearly two decades. To those who listened patiently and offered their sincere comments, I am most grateful. And to those who took time to read my writings, before and during the preparation of this book, I give my most humble thanks.

The evidence in this book came from many sources. They are the unsung heroes of any scientific advancement: those who carefully collect facts and prepare the textbooks. Liberal use has been made of a few excellent geology texts which represent the state of the art as it is being taught today. Howell's *Introduction to Geophysics*, Leet and Judson's *Physical Geology*, Dunbar and Waage's *Historical Geology* and Grabau's *Principles of Stratigraphy* are particularly clear and concise. Outstanding among forward-looking publications was the collection of undersea facts prepared by Dr Francis P. Shepard, whose book *The Earth Beneath the Sea* and prior papers about submarine canyons contributed greatly to both the development of the synthesis and the writing of this book.

Contents

Preface

The history of the world has been written many times by many authorities. Some have been transparently foolish; others have been obviously slanted to bolster one goal or another. A few have been serious attempts to understand the past by thorough analysis of all the facts currently available.

But new facts continue to become available, and often they show that the voids between evidence have previously been incorrectly bridged. Since 1945, Earth's surface has been mapped as it never was before; the floors of the oceans have been revealed in some detail for the first time.

Since 1960, the space around Earth has been filled with data-sensing satellites that have further analyzed Earth's form and composition. In that time probes have been sent to other planets, and humans have explored the surface of the Moon.

Since 1970, ships capable of drilling core samples from far under the floors of the deepest oceans have been ranging the seas. And special miniature submarines have made it possible for people to visit the sea floors two miles below the surface.

From this wealth of new information has emerged a vastly different synthesis of the Earth's history. Has a new and significant geological theory been constructed? Only time and the further excavation of places such as the sea floor will make that decision possible. But at the very least, the story which is presented here has been termed interesting and thought-provoking.

The central theme of this book is the announcement that Earth's planetoid-impact basins are still discernible – just as they are quite evident on the Moon and the other planets; that ice ages and their resulting effect on sea levels have been profoundly underestimated; and that the presentation of these discoveries as related events comprises a new synthesis of Earth's history.

The synthesis is well-founded and factual; however, its radical departure from past hypotheses stimulates many new speculations and conjectures. They are also presented here because they are interesting. Some of them may be found entirely wrong when they are thoroughly investigated.

But if the reader will be tolerant of any erroneous conjectures, it will be found that the synthesis is a sharp new tool for exploring all fields affected by the earth sciences, and it will not be substantially modified by future research.

1

The Revolution

'To treat your facts with imagination is one thing; to imagine your facts is quite another.'
– Burroughs

Velikovsky's Revelations

In his book *Earth in Upheaval,* Dr Immanuel Velikovsky joined a long parade of people – stretching back into time – who have proposed that catastrophes of titanic proportions have wracked the Earth. Unlike the rest, Velikovsky cited countless formal reports by accredited earth scientists which proved beyond the shadow of a doubt that catastrophes have played a crucial role in the shaping of the Earth as it is today. His thesis was like a bomb thrown into the hallowed halls of the geological uniformitarian purists. They have spent lifetimes working with the doctrine of absolute uniformity, assuming and trying to prove that every major alteration of Earth's surface took millions of years of slow change.

The result was outrageous but predictable. Had Velikovsky's scholarship been less than impeccable, they could have ignored his citations, knowing full well that they would swiftly fade from the scene. But his collection of facts was so powerful it could not be ignored. He was consequently attacked solely on the grounds of his hypotheses by those who stood to learn the most from his basic data.

In dealing with the history of the Earth, it is rare to find evidence that is conclusive. Nearly all phenomena are subject to multiple interpretations. Both Doctors Robert S.

Dietz and Velikovsky study the same Earth and the same evidence of its physical evolution. Yet, Dr Dietz proposes that the major features of the Earth were formed over periods of millions of years, by the continents floating and drifting on the surface of the Earth; Dr Velikovsky proposes that those same features were caused by a single catastrophe 3500 years ago.

Most earth scientists are suspicious of absolutist view-points and conduct their investigations in a more moderate vein. This vast audience probably deduces that although uniformity doubtless prevails for long periods of time, it has certainly been periodically interrupted by catastrophes that have had a devastating effect upon life forms and the Earth's surface. The synthesis presented here will attempt to disclose the primary mechanisms of the catastrophes so that they may now be identified and better understood.

Brooks' Discovery

In 1949, in his book *Climate Through the Ages*, C. E. P. Brooks announced an interesting discovery. In fact, he had made perhaps the most significant discovery that has been offered in this century, but he minimized its importance because the dogma of absolute uniformitarianism requires that such discoveries must be misinterpreted.

He had studied geological history and noted that every so often the Earth seemed to go through a period of convulsions during which great mountain ranges were thrust up, volcanos broke through Earth's crust and flows of lava flooded the land. He termed these episodes orogenic 'revolutions'. Using the uniformitarian geological clocks, he plotted the periodicity of these revolutions, and then superimposed the dates of the episodes of ice ages. He noted and charted the regularly coincident occurrence of these events. But using the uniformitarian geologic clocks led him to propose that each ice age consistently *followed* each revolution by a million years or so. This apparently vast spread of time between mountain building and ice sheets probably removed from his consideration any possible cause-and-effect relationship, except perhaps some vague long-period effect that mountains ultimately have on climate.

If, however, the evidence is studied without prejudice about 'clocks', a direct relationship may be seen. Physical evidence of an ice age consistently surrounds mountain-building episodes. That order of events supports the alternative hypothesis – that ice sheets cause mountains – instead of the opposite conclusion.

Brooks missed his opportunity to disclose this hypothesis, and also the derivable hypothesis that every prior estimate of the time between each pair of events was wrong by a factor of 'a million years or so'. On the other hand, had he thought of it, it is probable that he would not have published the idea, so great is the pressure of potential scorn upon those who dare to suggest that catastrophes have interrupted the slow-and-steady procession of Earth's events.

Here there is no such constraint, so this synthesis may give voice to many ideas that distinguished investigators have chosen to leave unspoken, and which in total show Earth's history to be enormously different from earlier proposals.

Old Facts, New Questions

At present, the absolute uniformitarian hypothesis with the widest acceptance is that of 'plate tectonics', also known as 'spreading sea floors' or 'drifting continents'. To support this hypothesis, its advocates have put their interpretation to many facts about Earth and the solar system. The best known of these is the fact that the eastern side of South America vaguely resembles the shape of the western side of Africa. They also make much of the fact that the mid-oceanic submarine ridges are locales of volcanic activity and of some degree of magnetic parallelism. But there are many other significant facts that do not fit their interpretations, and questions will be raised about these.

FACT: Every planet in the solar system has been found to

have internal heat and a magnetic field. What causes these?

FACT: The Earth's magnetic field fluctuates on a regular daily basis. It also fluctuates over longer periods. What causes these fluctuations?

FACT: Seismic studies of the Earth are compatible with either a hot or cold center. What are the implications of a cold center?

FACT: The Earth is covered with circular features, many of them hundreds of miles in diameter, at all altitudes. What caused the formation of these circular features?

A study of minerals and where they are found also leads to some puzzles. New thinking should be applied to the following:

FACT: Great placers of pure elements and refined compounds are not being formed by nature today, yet they exist in large quantities. How were they originally separated and deposited?

FACT: The miles-deep Gulf of Mexico and the Mediterranean Sea have buried floors of salt, gypsum and anhydrites. Have these bodies of water periodically dried up completely?

FACT: The present hypothesis of the formation of oil leads to success in drilling less than 10 percent of the time. How is oil formed, and where may it be found?

FACT: Cratons, the uncontorted 'plates' of the drifting continent hypothesis, contain vast sheets of coal. What was the source of pressure that converted these forests into anthracite and bituminous (hard and soft) coal?

FACT: Analysis of the surface of the Earth shows that much of its area is distributed between two flat platforms, one approximately at sea level and the other three miles below sea level, under the oceans. What formed these two platforms?

The sea floor, too, has yielded new facts that must be carefully considered in the light of their implications.

FACT: A system of submarine canyons greater than any on land leads from the upper platform to the lower platform, three miles under the ocean. How were these canyons eroded?

FACT: Corings of the sea floor show that deep-sea deposits alternate with layers of limestone, a rock that can only be formed in shallow seas. Has sea level undergone frequent three-mile changes?

Well-known facts about mountain ranges, too, must be reexamined.

FACT: Many ranges are made of folded rock. Was that rock hard when folded, or was it in a plastic form?

FACT: Man produces cements which form rock when mixed with water. Does nature also produce some rock in this manner?

FACT: Although mountain ranges may rise many miles above sea level, their heights are limited to about three miles above their formation-pressure sources. Why are they so limited?

The history of life on Earth raises questions which are not answered in conventional biology. Some interesting points are:

FACT: Although geneticists have never produced a new species of plant or animal, large numbers of new species have burst upon the continents many times in the past. Where have these new species suddenly come from?

FACT: Animals and plants have migrated freely among all the continents except Australia. How were such migrations possible, and why was Australia excluded from many of them?

FACT: Vast numbers and whole species of well-adapted animals have suddenly become extinct many times. Why has this happened?

All of these well-known facts are products of meticulous research, observation and investigation. Some are easily adapted to the drifting continents hypothesis, but others tend to refute the hypothesis. The synthesis presented in this book will relate these facts to each other in such a manner that these conflicts do not arise.

2

Life Meets the Catastrophes

Sudden Death

As Velikovsky so convincingly shows, a catastrophe of terrible proportions extinguished all the plant and animal life in Northern Siberia, Canada and Alaska at a single stroke. In these lands huge deposits of ground-up animal bodies lie mixed with splintered trees, soil and silt. North of Siberia are the 'Ivory Islands', so named because they appear to be made entirely of prehistoric mammoth bones and tusks. And in Siberia are the famous mammoths whose frozen meat was used for untold years as food for humans and animals. All around them were more thousands of skeletons of elephants, rhinos and other mammals of the period.

Nor is this deposit of smashed and broken bones in tremendous quantities unique. The Caloosahachee River in Florida flows from the famous Lake Okeechobee in the south central part of the state of Fort Myers on the Gulf of Mexico. Along this river are found remarkable bone deposits from the time of the last ice age. The uppermost shell beds contain vast quantities of marine invertebrates – all churned up by some force – and with them are the dismembered and shattered skeletons of a strange mixture of animals. These include the crocodile, alligator and horse. Elsewhere in Florida are found camels, tigers, mastodons – the complete spectrum of grazing and predatory life of the Pleistocene Epoch, both land and marine.

The animals in the Florida deposits lived at the same time as the mammoths of Alaska and Siberia, although the difference in climate which still prevails caused some difference in the types of fossils. But the profusion of fossils and the smashed, dismembered nature of the remains describes a tremendous catastrophe which snuffed out the lives of vast numbers of animals and deposited their remains in great piles.

Similar deposits of huge numbers of animal bones and forests of crushed and broken trees are found throughout the world. In the American western interior are deposits, also dating from the Pleistocene Epoch, which give mute evidence of a catastrophe that changed the face of the Earth. At Agate Spring Quarry in Nebraska there are tremendously thick deposits of disarticulated and broken bones. Again, oddly mixed and broken bones, including those of seven humans, were found in a deposit near Choukoutien, China. In India, in the foothills of the Himalayas, there is a rich fossil bed with many and varied types of animals, all apparently killed and buried at the same time.

At the other end of the Himalayas, in Burma, there is another great deposit of fossil animals and plants. These are particularly reminiscent of those of Alaska and Siberia. Besides the fossils, there are massive deposits of sand and sandstones and huge piles of petrified trees. The descriptions of these deposits match almost perfectly with the descriptions of the massive piles of petrified trees in Siberia.

But what is truly amazing about the deposits in Burma is that they lie at an altitude as high as 10,000 feet. The deposits include recently extinct or still extant species such as the mastodon, hippopotamus and ox. These animals were contemporary with modern man, and probably lived near sea level. But during the time that man has existed on Earth, those deposits have been elevated 10,000 feet. Compared with geology textbooks that assert that the process of mountain building takes place over millions of years, these mountains have sprung from the ground like a jack-in-the-box.

What catastrophe could have caused the death, burial and subsequent fossilization of these animals all over the world? All of these deposits were apparently made at the same time or during a very brief period. Animals were slaughtered at a sudden stroke, and their bodies were gathered up into huge piles. Mountains sprang up, elevating the bodies of animals native to the plains to high altitudes.

This catastrophe occurred within the time of human beings on Earth, as shown by the skeletons at Choukoutien and by the human tools that have been found among the deposits of mammoth bones in the frozen ground of Alaska.

In the riddle of the nature of the catastrophe that destroyed animals and mankind all over the world and ended all life in Northern Siberia, Canada and Alaska, nature has provided us with several excellent clues in the perfectly preserved frozen mammoths. Scientists examined such an animal in 1901, so well preserved that its meat was edible after thousands of years. They found that it had broken its ribs, hip and a shoulder blade, and had much clotted blood in the lungs and unswallowed food in its jaws. They concluded it had died a sudden and violent death. Since it was found in the remnants of an ice sheet, they decided that the animal had fallen into a crevasse and died struggling to extricate itself.

The analysis of the speed of the animal's death is good, but the opinion on the cause of death is deficient. The mammoth was perfectly preserved, indicating that it was frozen very quickly after death, before decomposition could begin. It could have fallen into an ice crevasse, but food was found in its mouth, and ample quantities of undigested food have been found in the stomachs of other frozen mammoths. It is difficult to imagine that large quantities of plant life existed under temperature conditions that would freeze a dead mammoth solid before decay began.

Suppose that, as the mammoths were feeding on the grasslands and meadows of a warmer Siberia, a cold and toxic wind came up, a wind unlike any that modern humans have ever experienced. Across Siberia and Alaska the wind rushed, killing all in its path. As the mammoths grazed, the wind hit like a hammer blow. With their first few breaths of this incredibly icy blast, the linings in the mammoths' lungs froze, and then cracked as they tried to breathe. The lungs began to hemorrhage, and swiftly became filled with clotted blood. The mammoths suffocated, drowned in their own blood. Later, most would be torn to pieces, but a few would remain intact, except for a few broken bones, until uncovered thousands of years later.

Investigators have long proposed that a volcanic explosion, pouring gases into near space, could have been the reason for the cold blast returning to quick-freeze these beasts. Clarence Birdseye, the man who developed the fast-freezing processes for the preservation of food, is reported to have addressed himself to the problem of the perfect preservation of the mammoths. After studying the undigested food in the stomachs of the mammoths, he proposed that an instantaneous exposure to a temperature no warmer than −150° F would be necessary to preserve the stomach contents in the unrotted state in which they were found.

As this is being written, other compelling evidence has appeared from an unexpected source. From Tennessee has come an agatized honeycomb. This author was shown the bees within it, which were turned to stone in various activities ranging from egg-cell dormancy to honeycomb repair. All who have seen it agree that it yields to no other explanation but gassing or superfast freezing, with subsequent deep burial before any thawing took place.

Some questions arise in analyzing the frozen and fossilized evidence and trying to account for it. First, is it necessary to assume that the mammoths, rhinoceroses and other animals found in the deposits of Alaska and the islands of the Arctic Sea were brought there from some distant and more hospitable homeland? The mammoths, although found 60 miles within the Arctic Circle, and more than 2000 miles north of the present range of living elephants, were quite at home there. Unlike all living elephants, they bore a thick coat of reddish-brown wool interspersed with long black contour hair. They were 'woolly' mammoths belonging to a now-extinct species that lived in the northern parts of Eurasia and North America before the last great ice age. They were as well adapted as today's reindeer is to the frigid climate. The rhino and lion remains found in these same deposits are also not those of our comparatively puny modern-day beasts. They were hairy, sturdy creatures well adapted to life near and within the Arctic Circle.

Was there sufficient vegetation in these areas to support the large numbers of animals whose remains have been frozen in such quantities before the last ice age? Today, farming takes place even in the northernmost parts of Siberia, and areas of Alaska are farmed commercially. The trees that make up the 'petrified wood hills' of the New Siberian Islands and the plants which fed the huge numbers of animals probably grew where those bones are now found.

These isolated instances of violent death of masses of animals lend themselves to two interpretations: the slow, non-catastrophic sequence espoused by the absolute uniformitarians; or the sudden change of living conditions with which the life forms could not cope.

Were there only these isolated instances, we would be compelled to give the alternative views equal weight. But such is not the case. These examples are very minor cases of massive extinctions that have completely wiped out many life forms over the entire planet suddenly – from the Poles to the Equator – and with a severity that prevented any of their seed from ever again living on the face of the Earth.

That *is* the definition of a catastrophe.

Flesh and Bone Turns to Stone

Thanks to the work of William Smith, a surveyor now heralded as the father of modern geology, for a century or more stratigraphers have been able to define different eras in the Earth's history according to the kind of life that existed. For example, during the Cretaceous Period of the Mesozoic Era, monstrous dinosaurs roamed the Earth in large numbers. Dinosaur bones in a given rock therefore place that rock in the sequence relative to other life forms living before or since.

At the end of the Cretaceous Period, the dinosaurs disappeared completely from the face of the Earth, and the Cenozoic Era, or Age of Mammals, began. What caused the eradication of the dinosaurs and the sharp division between the Age of Reptiles and the Age of Mammals? Can the answer be discerned in the fossil record?

There are several types of fossils, reflecting different methods of preservation. One type of fossilization is petrification; the carbon in the cells of an animal is replaced by minerals. Sometimes the entire structure of the animal is replaced. Another type of fossilization is distillation, in which the volatile elements evaporate, leaving behind a small deposit of the original carbon structure. Coal is formed by this process, as is petroleum. A third type of fossil is the mold, in which rock hardens around a plant or animal. Later, the organic materials disappear, leaving an empty space exactly the size and shape of the item originally entombed. From this mold, an exact cast of the fossilized plant or animal may be made. Fossils may also exist in the form of traces – the burrows, trails or tracks left by animals that once lived there.

To understand the process of fossilization, it is first necessary to consider the mobility of the animal or plant which has been fossilized. In the case of marine animals which live in a confined space, or that move so slowly as to make an escape from danger impossible, it is easy to see that slow-and-steady natural events could cause their death, burial and preservation – just as the uniformitarians propose. Today's seashore gives ample evidence that shells may be buried. If long-term wind and wave actions combined to keep them buried, it is possible that fossilization could result.

In the case of more mobile animals, the problem of fossilization is much more complicated. The profusion of mammal, dinosaur and fish fossils raises the question: If the dangers which these animals faced and which ultimately killed them were not catastrophic in nature, why did they not retreat from them? Perhaps there is a parallel to the case of the disappearing dinosaurs in the similar case of the extinction of the woolly mammoths, also apparently well suited to their environment. Perhaps toxic cold blasts killed them, too, with the continuing cold preserving their bodies until they were buried. The catastrophes which killed the mammoths and the dinosaurs were separated by a vast gulf of time, but their nature and effects seem to be just the same.

The eradication of nearly all life on land at the end of the Age of Reptiles is easily explained by a blast of cold gases, and their preservation can be accounted for by continuing low temperatures followed by the burial of their frozen bodies. But what would the effect of the cold blast be on the creatures of the sea, whose fossils are also found in such profusion in some spots?

In shallow waters, it is possible that the reason for the preservation of the fish is similar to that of the dinosaurs. Although the change in temperature would be much less rapid in water, the toxic gases and inexorably falling temperatures would kill all fish, so that none would be left to devour the remains of another. The low temperatures would inhibit putrefaction by microorganisms, preserving the bodies in 'cold storage' until burial by volcanic ash.

But what of the fish of the deep waters? A sudden wave of icy toxic gas would not be enough to kill all the fish in these heat-holding bodies of water. Some would survive the cold by seeking deeper water, and their companion scavengers would dispose of the evidence of the existence of their unfortunate fellows. The deposits near Orkney, Scotland, in what is called the Old Red Sandstone are ocean fish. Perhaps they were suffocated and buried by vast quantities of volcanic dusts and sands which sifted down into their habitat and preserved them through the ages.

Uniformitarians propose that rare natural conditions which are observable today can account for the formation of fossils. One of the conditions that is frequently cited is burial by sediment during the flood of a usually quiet river. Another is the presence of poison gases on the floors of some bodies of water, preventing the existence of predators, scavengers, worms and bacteria, which would otherwise destroy the bodies of dead animals.

These conclusions seem to be borne out by the fact that vast numbers of shells of microscopic and larger marine animals are present in the form of fossils, thus apparently establishing the importance of a marine environment. Sometimes the fossils of mammals, fish and other larger forms of life are found with them. Mud deposits, too, have turned up fossils of creatures as large as the dinosaurs. This fact would seem to support the idea of burial by natural flooding.

But, if it were true that conditions which are observable today could cause the formation of fossils, we should today be able to find locations where animals and plants of later ages are being turned into fossils. The fact is we cannot. The carcasses of the buffaloes which were slaughtered in uncounted numbers on the American plains two generations ago have disappeared, leaving hardly a trace. Their flesh was devoured by predators soon after death, and today even their bones and teeth have crumbled into dust. The fish that die in our streams, lakes and oceans rise to the surface, decompose and disappear without a trace. Nowhere do we find today the great piles of preserved dead fishes that other ages have left for us to study. Our river deltas show not a single one of the thousands of dogs and cats that drown in the river each year being turned into petrifacts.

The process by which the sands and dusts (which originally buried fossilized plants and animals) are lithified (turned into stone) is important to the long-term preservation of those fossil remains. Perhaps the secret of this process of lithification can be found in the volcanoes. Of the material blown from a volcano, we find that bombs and large fragments accumulate near the volcanic vent. Volcanic ash, on the other hand, is transported by the wind, and the finest particles often circulate the Earth as atmospheric dust. Such dust later falls, mixes with clay, silt, sand and gravel and produces mixed deposits of all kinds, such as shales, sandstones and conglomerates, through silicification.

Silicification is a natural process by which noncrystalline

materials, such as glassy volcanic dusts, turn into crystalline materials, such as the rocks quartz and chert, with the help of acids, alkalis and water. This silicification process is interesting because it is capable of turning unconsolidated materials, such as layers of sands, muds or clays, into solid rock, without the millions of years which uniformitarians usually allow for the metamorphosis of rock.

This suggests that perhaps fossils can be formed best when animals or plants are quickly buried in the presence of activated dust which a volcano ejects. The water in the animal or plant, or any other surrounding water, then begins the silicification process. The dust changes from glassy shards into crystals – from loose dust into solid rock – preserving each life form in stone. During the rapid deposition of volcanic dusts trees are petrified and the bones and tissues of animals are lithified. These conditions are rare today, and that is part of the reason no large fossil beds are forming today.

The idea that fossils are produced by forces present in the modern world is particularly hard to support in the case of tracks and trails of ancient animals which have been preserved in hard stone. A person who leaves a footprint in mud does not expect to see it the next month, much less have it be preserved forever in stone. Yet a deposit in the mountains of Switzerland preserves from Eocene times the feeding trails of a deep-ocean bottom-dwelling invertebrate. Dinosaur tracks from the Cretaceous Period are preserved in Dinosaur State Park near Rock Hill, Connecticut, and animal trails have been found in marine strata deposited in Nova Scotia during the Silurian Period. It is hard to imagine how these and many other similar trails could have been preserved if not by a process such as the freezing of the mud in which they were first made, or by its quick burial by volcanic dusts and the cementation of those dusts into rock.

When considered as a whole, the quantity of fossilized remains that exist on Earth is staggering. Rather than being rare and valuable, some forms of fossils are so common that we mine them routinely for the minerals they contain.

One type of fossil consumed in trainload lots is coal. Coal simply is not being produced by nature today; therefore, something very different about yesterday's environment must have caused millions of square miles of forests to be leveled and preserved. Another 'fossil' is the oil which we use by the shipload for fuels and lubricants. Although this fossil does not preserve the original form of the animals and plants which formed it, it does preserve the chemicals of their bodies in slightly altered form. As in the case of coal, nowhere is oil being produced today in any but laboratory-scale amounts.

Less well known are the fossils which have yielded up the phosphorus of their bones and teeth to enrich the soil. In one location in Florida, so many animals were killed and buried that millions of tons of their phosphorus have been mined in the form of phosphate rock. The vast quantity of their remains have caused the area to be called 'bone valley'.

Fossils are found in large quantities in many locations and from all the ages of life on Earth. Since they are apparently formed during times of extreme cold and great volcanic activity, each example of a substantial fossil bed indicates a great catastrophe on Earth. The many fossil deposits confirm that worldwide catastrophes have shaken this planet many times.

Darwin's Dilemma

In 1859 Darwin published *Origin of Species*, a work on the theory of evolution which sparked a controversy among scientists that has persisted to the present day. Darwin proposed the idea of 'natural selection' – a concept that has such intellectual neatness that it has been widely accepted, despite the fact that there is much evidence that it errs in its description of the process of evolution.

Darwin's theory suggests that new species develop and old species disappear over vast periods of time, and that effects of evolution and extinction take place in small steps as animals adapt to the rigors of survival in the world. But observation of actual occurrences in the fossil records show that this concept of slow change is false. New species arise suddenly, and in great groups, rather than one at a time. Extinctions occur abruptly in the fossil record.

In the Precambrian, only the simplest forms of life existed in the oceans of the Earth. But, although fossils were exceedingly rare in Precambrian rocks, they appeared suddenly and in great abundance in the rocks of the next layer, the Cambrian Period. The fact that they appear suddenly and in great abundance is highly significant. Any suggestion of swift change is antithetical to both the doctrine of uniformity and the Darwinian theory of evolution. Yet there is no other way to describe the beginning of life on this planet as it is recorded in the rocks.

Again, vast and sudden changes in the nature of animal life took place between the end of the Cambrian and the beginning of the Ordovician. The trilobites changed their forms completely. The Bryozoa appeared, not with a single form but with a host of genera and species. Mollusks and cephalopods suddenly exploded into many different varieties, with no slow evolution into different forms.

The Carboniferous Period is subdivided into two major parts: first the Mississippian System, and later the Pennsylvanian System. Between these two systems, another revolution in life forms took place. Insects, centipedes, spiders and scorpions began to crawl upon the Earth. Amphibians appeared on land, the first vertebrates to do so. Also, two small forms of reptiles, the first of their kind, came into existence.

Uniformitarians are loathe to admit that there is a possibility that these new forms of life came into existence suddenly. They believe that the sudden appearance of such a varied array of land animals in the Pennsylvanian rock implies that their ancestors must have existed in Mississippian time and that their apparent absence can only be due to the lack of favorable conditions for their preservation by fossilization.

If there was only one geological period where there appeared to be sudden evolution without apparent precedent, it would be easy to agree that, for some reason, conditions had prevented the preservation of precursors of the Pennsylvanian amphibians and reptiles. But such revolutions are common in the history of Earth as revealed by its rocks, so another explanation must be sought.

The evidence of the ancient fossils shows that new species have repeatedly burst into the world with great suddenness and in large groups. There have also been repeated mass extinctions which have destroyed many species of animals at one time. Charles Darwin was aware of these problems, and they sorely troubled him during his lifetime. He found the extinction of large, strong, apparently well-adapted animals to be inexplicable in the terms of his theory of evolution.

Neo-Darwinists have explained the sudden appearance and disappearance of species by saying that the fossil record is faulty. They explain that in those places where there appear to have been mass extinctions or explosions of new species, there are actually millions of years of sediments missing. They propose that if those sediments were present, they would show that the now-extinct species died out gradually, or that the species that seemed to develop suddenly actually came from a long line of previous similar forms.

This explanation fails, because even the addition of millions of years to the fossil records of evolution will not correct the basic problems of the Darwinian theory of evolution. The first major problem is that a well-adapted species will not become extinct merely from the passage of time. If it is well adapted, it will stay well adapted, and will survive. Also, despite the apparent reasonableness of the Darwinian theory, *no new species* of animal or plant has ever been developed by nature in the time of man's existence. Scientists can induce mutations, but they have never succeeded in producing a new species.

The fossil record shows still more sudden changes in the forms of life on Earth. The Mesozoic Era, also known as the Age of Reptiles, was the era of the dinosaurs. During this era, the reptiles grew in size and carved diverse ecological niches for themselves. Reptiles of all sizes, from the smallest herbivore to the giant carnivore *Tyrannosaurus rex*, which walked on its hind legs and stood over 20 feet tall, roamed the Earth. The huge *Diplodocus*, nearly 85 feet long, grazed on the plants of the ancient world. Other reptiles swam in the oceans and soared in the air. Truly, the domination of the Earth by the reptiles was complete.

What then, happened to the hegemony of the dinosaurs? Why are we not still menaced by the great carnivores, nor inconvenienced by *Brontosaurus* devouring crops or blocking highways? There seems to be no good answer to the question in textbooks. These large, well-developed and well-adapted creatures were wiped from the face of the Earth, to have all of their ecological niches assumed by other types of animals. In the often quoted words of George Simpson:

> The most puzzling event in the history of life on Earth is the change from the Mesozoic, Age of Reptiles, to the Age of Mammals. It is as if the curtain were rung down suddenly on a stage where all the leading roles were taken by reptiles, especially dinosaurs in great numbers and bewildering variety, and rose again immediately to reveal the same setting but an entirely new cast, a cast in which the dinosaurs do not appear at all, other reptiles are supernumeraries, and the leading parts are all played by mammals of sorts barely hinted at in the preceding acts.

Those who accept the doctrine of absolute uniformitarianism assume that the minor changes which are observable on Earth, if continued over vast periods of time, could account for all of the major alterations which have occurred on the Earth's surface. Charles Darwin, using the same type of reasoning, concluded that if the minor mutations which occur naturally today continued for millions of years, vast changes in the structure of animals and plants would occur, and new species would develop.

In fact, however, development of new species and extinctions of old ones occur suddenly and in groups, rather than singly. It is the catastrophes in Earth's past that have induced these sudden, widespread changes. Sudden ice ages cause the extermination of animals that are apparently well adapted to their climate and terrain. And who can say what stresses animals were subjected to in those times? Perhaps large amounts of toxic and radioactive material were belched from volcanoes. Perhaps dusts in the atmosphere caused toxic chemicals to permeate the water supply, the plants and even the animals that survive the onset of the catastrophe. Perhaps there were still other stresses present in the life of the animals during these catastrophes.

Regardless of the exact mechanics, it appears that catastrophes, not vast periods of time, have a fundamental influence on evolution. Catastrophes cause widespread extinctions of animals that are apparently fit. They also subject animals to stresses which cause them to develop new species. And, according to the fossil record, this all takes place in a very brief time.

3

Battles of the Extremists

Uniformity versus the Bible

The concept that underlies the drifting continents idea of the formation of the features of the Earth and the development of life upon it is the doctrine of uniformitarianism. Unqualified acceptance of this doctrine has seriously hampered the progress of geology. The doctrine of uniformity is stated: 'Present continuity implies the improbability of past catastrophism and violence of change, either in the lifeless or in the living world; moreover, we seek to interpret the changes and laws of past time through those which we observe at the present time.'

This doctrine, so reasonable at first blush, loses much of that reasonableness when considered carefully. The phrase, 'Present continuity implies the improbability of past catastrophism', essentially means that since modern humans have no incontrovertible record of catastrophes, we may choose to assume that none have ever occurred. Considering the age of the Earth and the danger of misinterpreting its record of events, it is hazardous to extrapolate on the whole history of the world from any absolutist viewpoint.

The uniformity concept achieved acceptance in the scientific world of the eighteenth and early nineteenth centuries because these were troublesome times in Europe. Revolutions and wars caused much disturbance in the minds of people, and theologians were insisting that geologists fit all fossil evidence into two catastrophic events – the Creation and Noah's Deluge. It is understandable then that when the hypothesis of uniformity appeared, an idea which said that the Earth had suffered no catastrophes and would suffer none, it was immediately welcomed.

One of the major stumbling blocks to the development of the science of geology has been the question of the age of the Earth. Theologians had placed the Creation at either 3928 B.C. or 4004 B.C. The pioneer geologists of the time worked under the handicap of battling theologians over the age of the Earth and dual-catastrophism at the same time that they were discovering the basics of their science.

The intellectual battle of Geology versus Theology was eventually won by the geologists; still, at the time of Lyell's birth, although the Creation in six days was regarded as metaphor, Noah's Deluge was still considered a part of the history of the Earth.

Lyell took up the newly respectable idea that the Earth was one million years old and developed it further. He concluded that, in fact, there had been no Deluge, and indeed no major catastrophes at all since the Earth was created. Lyell rejected the theory of catastrophism – not because he had found any new evidence to refute it, but because he disliked the idea of physical violence in any form. In his view, catastrophes were a shelter for scientists who failed to subscribe to his proposal that infinite time plus small forces exactly equaled sudden large forces. In fact, his rejection of catastrophes was an absolute position as much based on faith as was the theologians' belief in the Creation and Deluge.

A world in turmoil was happy to share this faith, however. Wars and revolutions might change the life of a European, but at least they didn't have to worry about more fire, brimstone or another Deluge.

Darwin's theory of evolution is based on the doctrine of uniformity. This being so, the number of families of animals should change in a uniform manner throughout the history of the Earth, if the doctrine of uniformity is correct. The number of families might expand, contract or remain the same, but whichever occurs, the doctrine of uniformity demands that the change be constant.

Studies of changes in the number of families of animals

show that this has not been the case, however. The overall trend has been in the direction of increasing numbers of families, as would be expected if the theory of natural selection is true. But there are wide variations from a constant, steady growth rate that tend to refute the idea that this has been a uniform process.

The Cambrian Period is the first time in the history of Earth which has left us an abundant fossil record. Throughout this period, the number of animal families increased slowly, from about 80 to about 110. During the Ordovician Period which followed, the growth rate accelerated at first and then tapered off, ending with about 225 families. Next came the Silurian, which shows the same growth pattern; a sudden burst of growth followed by a slowing down in that growth.

From the beginning of the Devonian to the middle of the Carboniferous Period, a slightly different and more surprising pattern emerges. The number of families grows gradually, from about 300 to a peak of about 340, but then there is an actual decline in the number of families to about 320. This decline is irreconcilable with the theory of uniformitarianism.

From the middle of the Carboniferous, through the Permian, to the middle of the Triassic Period, there is another sequence of growth and decline. The number of families grows slowly to about 340, and then begins a decline which accelerates until the number of families diminishes to 230, a level not seen since Silurian times. The number of families then resumes its growth through the rest of the Triassic and the Jurassic Period. At the beginning of the Cretaceous Period, the growth rate jumps. This rate of growth abates slightly, and the number of families then increases steadily to the approximately 900 families of animals found today.

The inconsistent growth and occasional actual decline in the number of families of animals clearly does not square with the doctrine of uniformity. Some external factor must have affected the process of natural selection to make it so unpredictable. This evidence was not lost on Darwin. He wrote:

> It is impossible to reflect on the changed state of the American continent without the deepest astonishment. Formerly it must have swarmed with great monsters; now we find mere pigmies, compared with the antecedent, allied races.
>
> The greater number, if not all, of these extinct quadrupeds lived at a late period, and were the contemporaries of most of the existing seashells. Since they lived, no very great change in the form of the land can have taken place. What, then, has exterminated so many species and whole genera? The mind at first is irresistibly hurried into the belief of some great catastrophe; but thus to destroy animals, both large and small, in Southern Patagonia, in Brazil, on the Cordillera of Peru, in North America up to Bering's Straits, we must shake the entire framework of the globe.

Indeed we must.

Darwin's faith in the doctrine of uniformity caused him to deny the evidence of his own eyes, and the faith of Lyell has blinded geologists right up to the present day. Catastrophes have in fact shaken the framework of the globe, time and time again, and the fossil evidence stretching into the dim past proves this. Catastrophes periodically have halted the growth of animal life and exterminated whole families of animals and plants; and, at the same time, catastrophes have formed and destroyed mountains throughout the world. Uniformity was a comfortable idea, but later it was to become a straitjacket to scientists seeking to better understand the history of life and the earth.

The Time Traps

Humans have always been fascinated with their own heritage and history. The question, 'Where did I come from?' has been answered in every culture. The most primitive of societies use myths and religious stories to explain their origins, while advanced cultures use science to determine the answer, but the question is still the same. Curiosity about our heritage exists in each of us, and the pursuit of the answer to the question of our origin is one of the characteristics that sets man apart from the other animals.

Humans not only want to know who their ancestors are, they want to know how far back he can trace them. Today in the United States, the ability to trace one's ancestors to the colonists that landed on the *Mayflower*, or to colonists during the Revolutionary War, is cause for admission into exclusive organizations. Some people can trace their family genealogy back for many generations and many hundreds of years.

Until 1749, western civilization based its concepts of the age of the world and humans on the writings of the Bible. Theologians taught that the Bible was literally true, and that the story of the Creation was a factual account of what had happened. At that time, many people had had their genealogies traced back to Adam and Eve. Biblical scholars calculated the Earth had been formed about 4000 years before the birth of Christ, and this was accepted as the Earth's actual age.

In 1749, the French scholar Comte de Buffon published the theory that the Biblical story of the Creation was metaphor, and that the six days mentioned were in fact six great epochs in the Earth's history. Religious authorities forced him to recant this view, but the revolution had begun. In 1785, the Scottish geologist James Hutton published *The Theory of the Earth*, proposing the doctrine of uniformity. When Charles Lyell took up the cudgels for uniformity, a new concept of the earth's age developed.

Perhaps the worst flaw in the doctrine of uniformity is its use of vast periods of time to explain changes that appear to be impossible under normal circumstances. The folding of rock and the appearance of new species of plants and animals, both effects that are impossible to produce in the laboratory, are supposedly possible over great periods of time, usually in the millions of years. Apparently, uniformitarians believe that if one waits long enough, the laws of physics are repealed and all manner of amazing things may occur.

The doctrine of uniformity dictates that in order for great changes in the Earth's surface to occur, endless periods of time are required. Unfortunately, this predisposes many people to believe any report that shows a fossil or geological formation to be millions of years old. Students of geology should beware of this trap, for it is based on circular logic: Changes in the Earth take millions of years and there appear to have been many changes, so the Earth is very old; also, current dating systems show the world to be very old, therefore the changes that have taken place have had millions of years in which to occur. Each assumption supports the other, but either assumption may in fact be faulty, making the other equally untrustworthy.

This book will not attempt to propose a new age for the Earth or an exact timetable for the growth of mountains and the alteration of species. It will, however, propose means by which these changes may take place in tens, hundreds or thousands, but not millions, of years. Perhaps with these means of change in mind, the geological clocks may be reset to a more realistic speed.

The Greek historian Herodotus presaged later attempts at dating in about 450 B.C. He observed that the Nile, during its annual flooding, spread a thin layer of sediment over its valley. He then determined from the thickness of the sediments in the Nile Valley that the valley and river were many thousands of years old. In 1861, two scientists used roughly the same method to determine that the Mississippi River was about 5000 years old.

The first serious, scientific attempt to determine the age of the Earth by use of the doctrine of uniformity was done in 1899 by John Joly. He attempted to show the age of the oceans from the amount of salt which they contained. In order to determine the age of the oceans, though, he had to make three assumptions: (1) The ancient oceans were not salty; (2) the salt that made those oceans salty came from the weathering of rocks and went into solution in the ocean; and (3) the rate at which salt is entering the oceans today is the same rate at which it has always done so. Using these assumptions, he determined that the oceans were approximately 100,000,000 years old.

But a theory can be no more reliable than the assumptions that it is based on, and the three assumptions that Joly used were all shaky. First, he did not know that the oceans were not already somewhat salty when they were formed. If there were any salt in the oceans when they were formed, Joly's method of calculation would make the oceans appear older than they actually are. Second, not all of the salt that enters the oceans is from the original decomposition of minerals. For example, some comes from salt mines and millions of tons of salt are mined each year. All of this salt ends up in the ocean, speeding up its apparent aging. Third, there is little reason to believe that the rate at which salt has entered the oceans has been constant. The great changes of world climate that have occurred have drastically altered rainfall and drainage patterns many times. The constantly changing climate makes it impossible to guess at what average rate salt has entered the oceans since the beginning of time.

Joly's method of determining the age of the oceans was 'scientifically valid' when that determination was made, since the doctrine of uniformity allowed that the forces that were acting on them then were the same forces that have acted on them since the beginning of time. Rivers, therefore, were dated by the amount of sediments in their deltas, glaciers were dated according to their greatest extent and their present rate of speed and mountains were dated on the basis of their rates of erosion. But all of these datings suffered from the same basic flaw: The forces that have acted upon the Earth have *not* been constant. Therefore, hypotheses based exclusively on the theory of uniformity could not provide reliable information on the Earth's early history.

Similarly, Lord Kelvin, the great British physicist, attempted to determine the age of the Earth by heat loss in a series of papers written between 1862 and 1897. He assumed that the Earth had begun as a molten ball and that it radiated its heat away into space over the millennia. His methods were impeccable, as one would expect, but his assumptions were faulty. He assumed that the Earth had begun its life as a molten mass of liquid rock, an idea which is losing favor in scientific circles. Also, he assumed that all of the heat which the Earth was radiating was left over from its formation. The discovery of radioactivity in 1895 made that assumption invalid, since the Earth's radioactivity is generating some new heat all the time.

Ever since the birth of the science of paleontology (the study of fossils), geologists have been able to arrange rocks and formations in the proper order, or sequence. The fact of evolution means that the fossils that are found in a certain layer of rock identify it as being different from other layers of rock. This, combined with the principle that newer layers of sediments are deposited on top of older ones makes possible quite accurate accounts of the order of events on the Earth.

After years of study and correlation of data from different locations on Earth, paleontologists have been able to produce a good picture of the effects of evolution. Therefore, if a geologist finds a fossil in a layer of sediment, he can determine from the nature of the fossil when in the history of the Earth that rock was laid down. By extrapolation, he can determine further information about the layers of rock above and below the fossil.

But this sort of dating only tells relative times, not absolute times. A geologist who finds a fossil fish in a layer of rock can tell when, during the evolution of fishes, that rock was laid down. It will tell him whether the rock is older or younger than other rocks found elsewhere. But paleontology alone cannot tell how old a rock is in years, and that fact has led to a continuing quest for something that will.

For the very recent past, tree rings provide that kind of absolute dating. Trees form one growth ring each year, so by counting growth rings backward from the year the tree was cut down, scientists can determine information about particular years. The bristlecone pine, a tree of the western United States that may live to be as old as 5000 years, is providing much information dating back as far as 6200 B.C. Unfortunately, these rings can only provide climate information in the changes of their structures and their rate of growth. Also, their range of time is very limited in relation to the Earth's long history.

Another means by which scientists have attempted to apply an absolute dating system to geological formations is by means of varves. A varve is a pair of thin sedimentary beds, one coarse and one fine. This couplet of beds is interpreted by uniformitarians as representing the deposits of a single year and is proposed to form in the following way. During the period of summer thaw, waters from a melting glacier carry large amounts of clay, fine sand and silt out into lakes along the ice margin. The coarser particles sink rapidly and blanket the lake floor with a thin layer of silt and silty sand. But as long as the lake is unfrozen, the wind creates currents strong enough to keep the finer clay particles in suspension. When the lake freezes over in the winter, these wind-generated currents cease, and the fine particles sink through the quiet water to the bottom, covering the coarser summer layer.

The Swedish geologist Baron de Geer related the varve deposits in many different lakes to determine that it took approximately 5000 years for the European ice sheet to recede from the southern tip of Sweden to the mountains in Sweden which still bear glaciers. Other scientists, extrapolating from his data, determined that the melting of the ice sheets began in Europe between 25,000 and 40,000 years ago, with the melting ending about 5000 years ago.

When more advanced radioactive-decay dating systems were applied to the problem of the ice age, however, quite different dates came up. This dating system indicated that there had in fact been an increase of ice 11,000 years ago, although the disappearance of the ice sheets coincided well.

The difference between 11,000 years ago and 25,000 years ago is considerable. But the system of varve counting is so simple and straightforward, it is difficult to imagine how it could be in error. The system is reminiscent of tree rings in its simplicity, and should be so in its reliability.

But let us take another look at varve counting, particularly in the hundreds-of-feet-thick Green River Shale formation of Wyoming. Within that formation, a fossilized fish about the size of your hand is sandwiched between many layers of varves. The layers above and below this fish are less than 0·007 inch thick. Assuming that the compaction rate for the sediments as they turn into rock is about 60 percent, this means that the layer of sediment that covered the fish the first year was about 0·016 inch thick.

One of the requirements of fossilization is that the animal or plant that is to be fossilized must be quickly buried to protect it from scavengers or decay by bacteria. But if the varve that covered our fish in the Green River Shale was the deposit of a full year, that means that the fish lay undisturbed on the bottom of a lake, with less than 0·016 inch of sediment covering it, and that this thin film of material protected it for a full year. After five years, this fish was covered with 0·08 inch of sediments, and still no scavenger ate it, nor did any bacteria attack and destroy it. Clearly, the fish could not be protected for this length of time by so thin a deposit.

So, perhaps the deposition of varves is not always an annual event. Perhaps these Green River Shale varves are the *daily* effects of a melting ice sheet. As the Sun warms the surface of the receding ice sheet, it melts at a prodigious rate, carrying the dusts and dirt on its surface into the nearest lake, stirring that body of water so that the dusts stay in suspension. But when night falls, the melting slows or stops. The lakes and rivers into which the day's meltwaters were pouring become still; the material that has been suspended by their motion falls to the bottom, coarse grains first, and fine grains on top of them. The Sun of the next day will bring a new rush of water and sediment.

If the varves which covered the fish in the Green River Shales were laid down on a daily basis, its preservation and fossilization become much more understandable. Under these circumstances, the fish would be buried by an inch of sediments in a little over two months. The icy waters of the melting continental ice sheets may well have preserved it long enough for that type of burial.

Viewing a varve as a daily phenomenon can also help bring Baron de Geer's schedule for the disappearance of the ice sheets better into line with the radioactive decay findings. The varve counting system had the ice age end at about the time that the radioactive-decay system said, but the process seemed to take too long. If the varve-making process is daily rather than annual, the findings of Baron de Geer should be divided by 365, and they may match better with the radioactive-decay reports.

How does this rate of deposition of material compare with the deposition rates observable in rivers today? In a section on reservoirs, the *Civil Engineers' Reference Book* reports that the reservoirs of the New River Water Co., London, England, were uncleaned for 100 years, during which time mud eight feet deep was deposited, or about one inch annually.

At Philadelphia, incoming water brings about a quarter-inch of silt in the reservoir per annum from the Schuylkill River, and one inch from the Delaware River. But at St Louis, Missouri, sediment brought in by the Mississippi River thickens the reservoir mud by about three to four feet per year. At a daily deposition rate of 0·016 inch, the Green River deposits were laid down at the rate of about six inches per year. Considering the enormous amount of water and glacial detritus that was carried away from the melting ice sheets, this rate of sedimentation is quite reasonable.

The field of atomic physics has given geologists a whole series of new tools with which to assign absolute ages to ancient materials. All of these methods are based on the fact that radioactive materials have a 'half-life'. Radioactive materials are essentially unstable, decaying over time into materials that are more stable. For example, the half-life of radium is 1620 years. If you refined a sample of pure radium, 1620 years later half of that sample would have decayed into other elements. Another 1620 years later, half of the remaining radium would have decayed, so that after 3240 years, only one-fourth of the atoms of the sample would still be radium. The other three-fourths of the material would have decayed into something else.

The beauty of dating by radioactive decay is that it is theoretically very reliable. The half-life of a radioactive material is apparently not alterable by any forces seen on earth, except in atomic piles. Therefore, if you found a sample of material which you know was once pure radium, which now consists of one-fourth radium and three-fourths the decay products of radium, you would be justified in saying that that sample was 3240 years old.

Scientists have used a number of different materials (called radioactive isotopes) in dating ancient substances. The first attempt at dating rocks by radioactive decay was made by the English physicist R. J. Strutt in 1910. He noted that the decay of uranium and thorium both gave off alpha particles. These particles captured electrons and thus became atoms of helium. Thus, he reasoned, by determining the proportion of uranium and thorium to helium in a sample, he could determine the age of the rock. The extremely high ages which he found for rocks, however, indicated that some of the helium had leaked away into the atmosphere. The leakage of helium made accurate dating by this method impossible.

Other work along the same lines was done at about the same time by B. B. Boltwood of Yale. Rather than consider the amount of helium given off by decaying uranium and thorium, he looked at the amount of lead that was left. Since uranium and thorium leave lead as their stable end product, measuring the proportions of uranium to lead will, in theory, give an accurate age for the rock sample in question.

A refinement to this method of dating rocks was later added through mass spectrometry by A. O. Nier. He found that uranium and thorium isotopes decayed into particular types of lead isotopes. Thus, uranium-235 decayed into lead-207, uranium-238 decayed into lead-206 and thorium-232 decayed into lead-208. Each of these types of decay could be used to make a separate measurement of the age of a rock, and in a rock which contains all of them (as many do) three separate dating processes may be used, with each useful as a check on the other two.

Besides the uranium/lead and thorium/lead methods of dating, several other types of dating methods are used by scientists. One of these is the lead/alpha particle ratio in zircon. The rate at which alpha particles are emitted from a sample can be used as a measure of age if the amount of radioactive and formerly radioactive material in the sample is known. So, to determine the age of a sample of zircon, the amount of lead in the sample is measured, and then the rate of emission of alpha particles is determined. One may then calculate the age of the sample.

Another method of radioactive-decay dating is the potassium/argon method. This is based on the decay of radioactive potassium to argon gas, and it determines age in much the same manner as the uranium/lead method. Also similar to these is dating by rubidium/strontium ratio.

One of the most interesting methods of radioactive-decay dating is that which was reported by Dr W. F. Libby of the University of Chicago chemistry department in 1948. This is known as the carbon-14 method. In the outer atmosphere, atoms of normal nitrogen are turned into atoms of radioactive carbon by the bombardment of cosmic rays. This carbon joins with oxygen to become carbon dioxide. In recent years, the proportion of this radioactive carbon dioxide in the atmosphere has remained fairly constant.

Plants use the radioactive carbon dioxide in photosynthesis just as they do normal carbon dioxide. As a result, the ratio of radioactive to nonradioactive carbon in living plants is the same as the proportion of radioactive to nonradioactive carbon dioxide in the air. Animals, too, have about the same proportion of radioactive carbon in their bodies since they acquire that carbon, directly or indirectly, from plants.

When a plant or animal dies, however, it ceases taking in new radioactive carbon. As with all radioactive isotopes, the radioactive carbon decays according to a predictable schedule. Therefore, if the amount of radioactive carbon in the tissues of the plant or animal at the time of its death is known, the age of the sample can be determined by the amount of carbon that is left.

The radioactive-decay system of dating is not without its assumptions, and these are not unassailable. In determining age according to the carbon-14 method, it is necessary to know the amount of carbon-14 that was in the tissues of the plant or animal at the time it died. Scientists do this by assuming that the amount of carbon-14 in the atmosphere has remained constant since ancient times, and that the amount of carbon-14 in the bodies of ancient forms of life is the same as that in living things today.

But carbon-14 is formed in the upper atmosphere of the Earth by bombardment of cosmic rays. If anything were to shield the atmosphere from those rays, the amount of carbon-14 in the atmosphere – and thus in the tissues of living things – would decline. In fact, a report on the bristlecone pine said that radiocarbon readings of bristlecone-pine rings, year by year, showed up errors in the conventional radiocarbon dating system that were caused by unsuspected fluctuations in the amount of carbon in our atmosphere . . . as a result, it was found that the time relationship between Europe and the Mideast was not all what had been supposed. It now appears that the great stone tombs in northwestern France and in Spain may be older than Egypt's famous pyramids – a startling thought. England's mysterious Stonehenge may be older than similar works elsewhere. Significant innovations in engineering, construction and metallurgy may have originated in Europe, not always somewhere to the east.

Even among those who have no theoretical objection to the idea of radioactive-decay dating, there are protests about its reliability. The extremely fine measurements that must be made, combined with the ever-present danger of contamination of sample or the results by stray radiation, is highly objectionable to some. These difficulties, on top of the problem of having no idea whatever of how much lead/isotope was in a sample when it was formed, make the results of radioactive-decay dating highly suspect.

Nearly all of the dating methods that were devised have been faulty, not because of faulty conception or sloppy application, but because they were based on unwarranted assumptions. They have assumed that the ancient seas were not salty, that all the Earth's heat was left over from its creation, that varves were an annual phenomenon and so on. Each of these assumptions effects the foundation of geology-related research, and many have been found defective. Since the latest assumptions may also prove defective, the Earth in fact remains undated.

Uniformitarians propose astonishingly long periods of time for the age of the Earth and the formation of its features in order to make up for deficiencies in the hypotheses regarding drifting continents and evolution. Dr Velikovsky compresses time to indicate that a great number of changes of the Earth took place during a short period 3500 years ago, to support the idea of catastrophes during Biblical times.

To cling to either of these positions exclusively is to halt investigation into the field of geochronology. Fossil evidence shows that there have been long periods during which species proliferated and were refined. But there have been catastrophic extinctions and sudden appearances of new species as well. Earth's surface too seems to have lain unchanged for periods of time, and then to have been contorted by periods of mountain building. A perfect means of establishing prehistoric dates may never be found. In the meantime, it is advisable to reserve judgement as to how long ago the ice ages displaced humans or whether they extinguished the dinosaurs.

4

Setting the Stage

This Dynamic Earth

Ordinary materials, such as rocks, are capable of being weakly magnetized. At many places around the Earth, there are rocks which have a magnetic orientation that is different from that of the Earth's magnetic field as it is today. This indicates that many times during its history, the Earth has had its magnetic field realigned. Uniformitarians place great store in these magnetic realignments as being solid evidence of slow and steady changes in the past.

The Earth's magnetic field obviously is not a changeless, immutable constant. The magnetic field changes regularly, to a considerable extent, and in a predictable manner. It is well known that the Earth's magnetic field has periodic daily variations. The daily changes follow the Sun, being generally similar for a given local time. Their connection with solar radiation is further shown by the contrast in activity between the changing daytime values and the more steady nighttime conditions.

Today, 6 percent of the Earth's magnetic field is externally induced by currents in the Earth and above it. That fact, combined with the fact that the rotation of the Earth in relation to the Sun alters the Earth's magnetic field, suggests that the relationship of the Sun and the Earth may be more complex than it first appears.

Besides the air currents and oceanic currents that are easy to observe, the Earth also has concentric shells of electrical currents. The most spectacular manifestations of these currents are observable during earthquakes. The night the

1930 Idu, Japan, earthquake hit, peculiar lights were reported in the sky in the vicinity of the epicenter. At places they were bright enough to illuminate objects. The color ranged from blue to reddish yellow. They appeared as radiating rays and fireballs. A similar nighttime discharge was observed in Wrightwood, California, in 1948. The lights in the sky that accompanied that earthquake resembled reflections of a welder's brilliant arc, except that the sky continued to glow after each initial flash for as much as 10 seconds. Each tremor was accompanied by a flash, and each flash reinforced the brightness of the sky during the minutes that the earthquake lasted. The illumination was obviously the result of electrical discharges into atmosphere that strongly ionized the particles and made them fluoresce.

Earth also has magnetic storms that show the intensity of the electrical surges that it sustains. In the early days of electrical communication – before microwave – these tidal waves of ions readily disrupted radio and teletype messages for hours at a time. The coincidence with sunspot activity led to the discovery that solar radiation of a corpuscular or electromagnetic nature influences the Earth's ionosphere, which results in electrical currents that produce magnetic disturbances in Earth's field.

The connection between magnetic storms on the Sun and increased aurora activity has also been established. And drastic weather changes on Earth can now be accurately predicted by astronomers noting the source of the disturbance, the solar flares.

Throughout these examples of electrical activity, the fact of changing magnetic fields is apparent. But what is the relationship between magnetism and electricity? In the most general of terms, electricity and magnetism are inseparable. If a wire is moved through a magnetic field, a current will be generated within that wire. If electricity is sent through a wire, that wire will form a magnetic field around itself. These two facts long ago led to the invention of the generator and the electric motor. The electric motor first transforms electricity into magnetism. It then transforms that magnetism into motion. The generator changes motion into magnetism, and magnetism into electricity.

Of most interest is the fact that both electric motor and generator are fundamentally the same machine. A generator may be turned into a motor simply by feeding current into it, rather than taking current from it. Similarly, an electric motor can be turned into a generator simply by turning its shaft.

Perhaps the relationship of the Sun and the planets is similar to that of the generator and motor. The Sun has a strong magnetic field, and sends out a solar 'wind' that bears a close resemblance to an electrical current. As the Earth moves through these on its orbit, with additional motion in the form of its rotation on its axis, both electrical currents and magnetic fields are generated in the Earth.

The origin of the Earth's 'permanent' magnetic field has puzzled some, but thinking of the Earth as similar to magnetically susceptible steel or ceramic helps to dispel this problem. The Earth can conduct electricity, and the motion of the Earth through the Sun's magnetic field certainly would induce the formation of electrical currents in the Earth. These electrical currents cause the Earth to become, to some degree, a huge electromagnet.

When an electromagnet has a core of steel or ceramic, the core soon becomes a permanent magnet.

The uniformitarianism hypothesis of the Earth's magnetism concludes that circulation of molten rock beneath the surface generates the Earth's magnetic field. Many materials may be magnetized, but all lose that capability when heated above a certain temperature, called the Curie point, which is well below the melting point. If the material under the surface is hot enough to be molten and flow, then it is also above the Curie point and cannot be magnetized. If it is cool enough to be magnetized, it cannot flow.

But the Earth is magnetized, and 94 percent of that magnetism appears to be permanent. Therefore, we may even speculate that the core of the Earth is cool and solid. Only then could it be permanently magnetized. Like so many other facts, the existence of the Earth's magnetic field jeopardizes the idea that the Earth is a molten mass with the continents floating on it. Heat is associated with electrical currents, however, and later we'll get back to that as a possible source of Earth's heat.

Rocks that have reversed magnetic orientation may, in fact, reflect reversals of the magnetic field of the Earth. But reversed magnetization may also be a result of the tendency of a magnetic field to induce an opposing field. Investigators have observed a systematic reversal of the magnetization of certain Adirondack Mountain rocks related to their iron content. In the laboratory it has also been found that some compounds, when cooled through the Curie point in the Earth's field, had remanent magnetism opposite to it.

There are those that believe the reversed magnetization of rocks means that the Earth's axis of rotation was changed by some outside force. Instead, consider the forces normally at work as the Earth travels through the electromagnetic output of the Sun at 18 miles per *second*, which is thousands of times faster than any manufactured generator armature can spin.

As would be expected of any conductor traveling through a magnetic field, the motion of the Earth causes electrical currents to flow within the Earth. These currents in turn cause a magnetic field to form around the Earth. But what would happen if the atmosphere were suddenly filled with particles of iron-laden dusts?

As the Earth continued to move through the Sun's magnetic field, electrical currents would be induced – above the Earth's surface – in the dust-charged atmosphere. The field generated by that flow would also induce magnetism in the Earth, but in an opposite direction. To observers on the Earth, the north and south magnetic poles would appear to have changed places. And with the thick dust in the sky obscuring the Sun, we would have no way of observing that a change of the geographical poles did not take place. The dust that engenders this pole reversal might be of cometary or volcanic origin, but it is not necessary that the axis of the Earth be tilted to reverse the magnetic field.

If the Sun's influence on Earth's motion can stir up the magnetosphere, why not the portion of the atmosphere that includes the weather – and the oceans?

The presence of jet streams in the atmosphere is very difficult to explain in terms of 'hot air rising – cold air descending – and momentum generated by Earth's rotation'. That's why diagrams of the atmosphere showing these forces at work always make the air layer appear to be thousands of miles thick above Earth's surface.

But it isn't all that difficult. As Earth moves through the solar wind, electrical currents are generated in both the oceans and the atmosphere. They tend to follow the Sun, which is apparently moving westward at the rate of 1000 miles per hour. Fortunately for us, permanent magnetic fields, fluid viscosity and thermal interferences all combine to prevent thousand-mile-per-hour winds from developing. Instead, much slower winds develop near the surface of the Earth, and even they are diverted from their westward path by eddy-current 'highs' and 'lows'.

At high altitudes, though, a different effect occurs. Since the atmosphere cannot keep up with the Sun, electrical imbalances are created between day and night hemispheres. These imbalances are evened out by highspeed flows of ionized air masses. The flows, which we call jet streams, neutralize the imbalances.

Similar electromagnetic stirring may cause the much slower currents in the oceans. These flows, such as the Gulf Stream and the Humboldt Current, are similar to atmospheric highs. The electrical nature of the Gulf Stream was pointed up by the University of Miami's School of Marine and Atmospheric Sciences. Researchers there found that the Gulf Stream appears to be a huge electrical generator that produces a measurable voltage; a finding that may be visualized in reverse, as shown by the generator-into-motor analogy.

The direction of rotation of eddy-current highs and lows

and the movement of jet streams in Earth's northern and southern magnetic hemispheres match precisely the movements expected of ionized-particle flows in such magnetic fields in the laboratory. This similarity alone is compelling evidence of electromagnetic forces at work on Earth's surface. It is therefore not surprising to find that tornadoes are now known to be sustained lightning bolts, and that hurricanes and typhoons on Earth are mirror images of sunspots. Hurricanes are born near latitude 8 degrees north or south of the Equator. They move away from the Equator and westward for considerable distances. At latitudes near 35 degrees, most of them lose power and dissipate. Sunspots are now known to be magnetic storms. They begin at about 35 degrees north or south of the Sun's equator and move eastward and toward it. When they are within about 8 degrees of the equator, they disappear.

The Electromagnetics of the Solar System

There are those who believe that the inclined orbit of the Moon proves that the axis of the Earth has been shifted. They cite the tidal hypothesis of solar system origin, which assumes that the plane of the Earth's orbit, the plane of the Equator and plane of the Moon's orbit should coincide. Since they do not coincide, supporters of the tidal theory conclude that the axes of the Earth's and Moon's orbits have been tilted.

In the past, many have also proposed that the Moon issued from the area of the Earth's Pacific basin, as the result of some cosmic disturbance. Recent lunar exploration has shown that the Moon was never a part of the Earth, but formed separately as a satellite of the Earth, probably from the primeval dust cloud that formed the solar system.

The dust-cloud hypothesis suggests that the solar system was once a formless cloud. Gravitational attraction among the particles caused the cloud to shrink, and vortex forces caused the cloud to spin. As the cloud rotated, centrifugal force caused it to spread from its 'equator' and to flatten into a rotating disk about the diameter of the present solar system.

The rotating disk divided itself into eleven 'shells' or concentric bands of material rotating at different speeds. The innermost band rotated the most quickly and later became the Sun. The other ten bands became the planets and the asteroid belt, with each band rotating at a lower speed than the one inside it.

This hypothesis would explain a solar system with ten planets, all of whose equators coincided with their plane of rotation, and none of which had any satellites. But the planets of the solar system have many satellites; and their axes are tilted at many different angles to the planes of their orbits; and the momentum division is all wrong.

To rectify some of the inconsistencies between the dust-cloud hypothesis and the observed facts of the solar system, let's add astronomer Fred Hoyle's magnetohydrodynamics (MHD), the forces of magnetism and electrical current, to the whirling cloud.

Thus, within the cloud, electrostatically charged particles were attracted to or repelled by each other. Because of their random motions, the particles that were attracted to each other rarely crashed. Instead, they approached close enough to alter each other's courses and then began to orbit each other. On a large scale, this swarm of minuscule orbiting charged particles created swirling electromagnetic fields throughout the cloud, as it spread and flattened into a rotating disk.

Also, each particle had a line of electromagnetic attraction, through other particles, which led to the center. Since the inner parts of the disk rotated at a higher rate than the outer parts, these electromagnetic lines became stretched. It was as if each particle were attached to the center of the disk by a rubber band. The center of the disk wrapped the line of force around itself and slowed its rotation by slightly speeding up the entire mass.

This constant winding and wrapping of the lines of force caused the disk to break up into eleven rings that later made up the Sun and the planets. But, since these rings were made up of moving masses of charged particles, each generated an electrical current surrounding its mass. Each of these rotating currents produced its own magnetic field, and since they were all moving in the same direction about the same point, there was serious conflict among the growing clumps which disturbed their motions. As a result, the areas of interface between rings were areas of great electromagnetic turbulence. The irregular flows of the charged particles in these interference areas produced magnetic fields that disturbed their orbits. Because of these irregular magnetic forces, as the bodies were condensing into their present forms, some blobs of material became satellites and many took on different inclinations and rotations. Perhaps this is why the Earth, instead of having its equator coincide with its plane of orbit, is inclined at an angle of 23 degrees and has a moon with a slightly tilted orbit.

Solar System Explosion Basins

Those who believe that catastrophes of global proportions have never occurred need only look elsewhere in our own solar system to be shaken in their beliefs. Between the orbits of Mars and Jupiter lies the asteroid belt. This belt consists of many small bodies, from less than 5 to about 500 miles in diameter, each of which revolves around the Sun in this collective orbit. The consensus is that the asteroids are fragments of a planet that was disrupted early in the history of the solar system, perhaps as it passed and repassed Jupiter.

Therefore, the Sun was once orbited by ten planets, but one of them was completely destroyed in a manner that we can only guess at. But the forces that acted on Aster (a good name for it?) can still be seen at work in the solar system. All planets are subject to many forces which affect them in various ways. Gravitational attraction tends to pull planets toward the Sun and each other; and centrifugal force tends to tear them away. The planets rotate about their axes and revolve around the Sun, their motions maintained by momentum, but the energy of that momentum is constantly being electromagnetically sapped and turned into heat. The planets also have chemical and radioactive sources of heat, and each planet receives a daily dose of surface heat from the Sun.

The nine surviving planets of the solar system have so far been able to withstand the combination of stresses that they encounter on their paths, but Aster was apparently larger and less stable. In an orbit next to Jupiter, now the largest planet in the solar system, it was subjected to greater stresses than were the other planets. The gravitational attraction between Jupiter and Aster perturbed Aster's orbit and put stress on the planet's structure. Perhaps the immense magnetic fields of Jupiter and Aster electromagnetically raised their internal temperatures. Then, under the tremendous thermal and gravitational forces, radioactive fission-fusion began within Aster and it exploded, scattering its broken pieces throughout the solar system.

The same forces that heated and tore apart Aster are working on the other planets all the time – but Aster had the misfortune to be large and close to Jupiter. Perhaps the heat induced in the Earth by electromagnetic eddy currents is enough to contribute to periodic violent volcanic activity and to account for the sustained internal heat; but fortunately for us, Earth is not massive enough to build up fission-fusion effects and explode like a giant atomic bomb.

However it occurred, the explosion of Aster had a profound effect on the other planets. It blasted chunks throughout the solar system, most of which collided with other planets. These pieces, ranging in size from planetoid to dust particle, left craters on all the planets we have been able to examine, and probably the others as well.

During the last few years, the formerly mysterious surfaces of Mercury, Venus and Mars have been disclosed by the great advances in space technology. The one factor which all had in common, to the amazement of many, was a profusion of craters. All resembled the pockmarked surface of the Earth's Moon.

Because of its thick cloud cover, Venus could not be photographed by probes, as were the others. But in 1973, the Jet Propulsion Laboratory of the California Institute of Technology used high-resolution radar to examine its surface. The radar, and subsequent computer clarification, disclosed that Venus was apparently as much scarred with craters as is the Moon.

Mercury, the nearest planet to the Sun, also has a surface that is riddled with craters. The pictures sent back by *Mariner 10* resemble exactly the surface of the Moon. Craters are everywhere, and some are hundreds of miles across.

But the planet that has received the most attention over the years has been Mars. Ever since the invention of the telescope, astronomers have peered at that planet, attempting to define its surface. For hundreds of years, people have traced the seasons on Mars and speculated on the 'canals' that seem to cross its surface. When Mars was finally photographed in detail by the Mariner Space probe, many were amazed to see that Mars, too, had craters. Rather than resembling the weather-smoothed Earth, as the poets had hoped for years, its surface was rugged, barren and scarred. Like the Moon, Mercury and Venus, its surface was covered with craters everywhere.

Finding craters on the other planets of the solar system is something of a letdown. We have been aware of these, all of our lives, on our own Moon. Those features have been mapped for hundreds of years, photographed since the invention of the camera and even visited by astronauts. But the presence of craters on other planets, too, forces a rethinking of the process of crater formation. Uniformitarians had assumed that the Moon was a special case. Their ideas of the formation of the solar system held that the Moon had been formed out of molten material at about the same time as the planets; but, being smaller, the Moon cooled and developed a hard surface quickly. Over the billions of years of the Moon's history, they proposed, it was inevitable that the Moon would get hit by larger numbers of meteorites. Lacking the atmosphere of a planet, the Moon had no protection from these meteorites, and it was hit by uncountable numbers of them over the vast amount of time. And since the Moon had hardened before Earth, it continued to bear the scars of those encounters.

But the discovery of craters on the other planets makes this line of reasoning untenable. Venus and Mars both have atmospheres, yet these atmospheres did not protect them from the impact of meteorites. In fact, the atmosphere of Venus is fifteen times more dense than that of Earth and still it has impact scars. Venus is only slightly smaller than Earth, with a mass four-fifths that of our world. Had it been molten at its creation, it would have simply swallowed up those meteorites without a trace. The similarity of Venus and Earth, and the fact that Venus is closer to the Sun and has a much hotter atmosphere, suggests that both planets should have cooled from this hypothetical molten state at about the same time.

All this leads to the conclusion that the Earth, too, once looked like Mercury, Venus, Mars and the Moon; but have our ancient craters all been destroyed by erosion?

The explosion of Aster bombarded the planets with huge planetoids traveling at tremendous speeds. Ordinary meteorites crash into the ground at relatively low speeds. They shatter, and splash material up to form deep, steep-sided craters, no more than a few miles in diameter. These rims are quickly obliterated by erosion.

The planetoids ejected by the explosion of Aster were very different from the meteorites. They were much larger, some of them hundreds of miles in diameter. They reached Earth at much higher speeds. As a result of their size and speed, these planetoids carried tremendous energy in the form of momentum. When they hit a planet, that energy was instantaneously converted to heat and pressure. The planetoids vaporized themselves along with the material in the area which they struck.

Instead of just splashing material into a steep crater, the impact of the high-speed planetoids caused earth-shaking explosions. A shock wave of surface material went out from the point of impact, sometimes for hundreds of miles. Eventually, the energy of the explosion was absorbed by friction, and the shock wave ceased its outward motion. But because of that friction, instead of the wave disappearing, as would a wave in water, the crustal wave was 'frozen'. It remained as a raised ring circling the point of impact. Within that ring lay a nearly flat plain, all of its features destroyed by the shock wave that crossed it. These very large, circular formations are here called 'explosion basins', to differentiate them from smaller 'craters'.

One look at a map or photograph of the Moon will show the difference between an ordinary crater and an explosion basin. The Moon is completey covered with many small, steep-sided craters. But the Moon also has many large, smooth, circular features, some of which are dark in appearance and are called *mares* (Latin for seas). These circular features, which are quite flat, are the explosion basins. Most of these basins have a few superficial craters which were formed after the basins themselves, but on the whole they are quite smooth and their circular nature is easy to see.

It is important to realize just how subtle the contours of an explosion basin are. Although they may be many hundreds of miles across, the highest part of the basin rim may be less than five miles above the lowest part of the basin. The rim of the basin, instead of being a steep formation like the ordinary craters, may be a broad rise that is miles across but only a few thousand feet elevated above the ground that surrounds it. To an observer on the surface, the basin is so large and its inclines so slight that it might be undetectable. But when viewed from a great distance, as we view the Moon, its circularity is unmistakable.

On Earth, the problem of locating explosion basins is further complicated by the fact that all land surfaces are constantly being eroded by wind and water. Mercury and the Moon at present have no atmosphere, so that erosion of their explosion basins by these forces is impossible. Venus and Mars have atmospheres; but little liquid water, because

Venus is too hot and Mars is too cold. On these planets, wind may erode the explosion basins, but the much greater eroding power of water does not threaten them.

Earth, however, has huge quantities of liquid water. That water covers most of the Earth, and is evaporated and rained down on the continents in a never-ending cycle. Water surf-lines constantly erode highlands and deposit materials in low areas. Thus, by erosion, the rims of Earth's explosion basins have been worn down and their enclosed basins filled up with sediments, thereby softening the already subtle contours of the explosion basin rims.

The idea that Earth's continents were formed by collision of other bodies with the Earth has long been discussed. If a great group of planetoids fell somewhere near the North Pole, splattering southward, it would adequately explain the limited distribution of the continents on the globe's surface, their elevation and composition and the way they taper in the southern hemisphere.

With all of the preceding in mind, consider this idea. Following the impacts that made the continents and the planetoid explosion basins, it seems likely that volcanic activity would cook fluids and gases out of Earth's interior to form its oceans and atmosphere. As water covered the Earth, those explosion basins above sea level filled with water whose surf action completely destroyed their rims. Some material was carried down outside the rim, and other material was carried to the interior of the explosion basin, filling and leveling it. Eventually, this erosion of the rim and filling of the interior produced the continental basins.

Those basins that were partially or completely below sea level are less rapidly being filled by debris, and thus have retained more of their original contours. So it may still be possible to recognize the larger ancient explosion basins, despite the deformation that they have undergone.

Since no other formation hypothesis of the features of the Earth proposes a mechanism by which circles many hundreds of miles in diameter can be formed, it is reasonable to suspect that any large circular forms found on the Earth are old explosion basins.

The search for such forms is complicated by the several types of modification that the basins have undergone during their time on Earth, and this makes it more difficult to see them than on the Moon and other planets. As mentioned before, those basins that were formed on the continental surfaces tended to erode to circular plateaus. But some of those plateaus became hosts to ice sheets, and those ice sheets have obliterated some rims. The circular forms of the explosion basins are also subject to another form of distortion – that of the mapmakers. The Earth is a globe, and in trying to draw a globe on a flat piece of paper, circles may become ovals, or even egg-shaped.

Also, because the planetoids usually landed at some angle, the rims they left are higher on one side than on the other; therefore, we may find the higher half of the rim only.

Despite all this, many have been found, and there is a remarkable example of one that is nearly a full circle. It appears on marine charts as the half-submerged North American Basin rim. This basin, under the western half of the North Atlantic Ocean, is 1800 miles in diameter. One rim of it is the northeastern coast of the United States and Canada, from Nova Scotia to Cape Hatteras. There the rim submerges and becomes the outer edge of the Blake Plateau and the Bahama Banks. Curving southeastward, it touches on the island chain that includes Puerto Rico and the Virgin Islands. From this point to the submerged Mid-Atlantic Ridge, erosion has obliterated the rim, but the Ridge itself is part of the rim, and it bends to indicate its relationship to this great circle. Finally we complete the circle with the famous fishing grounds of the New Englanders, the Grand Banks.

More and smaller circles will be identified when submarine contour maps are declassified, but for the moment, intensive search for such circles can be made on the well-mapped surfaces above sea level. The evidence of explosion basins is literally everywhere. The curving mountain chains and island arcs that compose the Earth's most prominent features have been long recognized as 'Earth's primary arcs', the name given to them by geologists who failed to see that some were of explosion basin rims. These sections of the rims have been covered by sediments that were later raised into mountains, making their identification difficult, of course.

It would be tedious to list them all, but more than 200 explosion basins with diameters over 240 miles have been identified on Earth's surface using ordinary commercial maps and charts. Perhaps the best illustration of such a basin is one that has its rim just above present sea level: the Gulf of Mexico. From the Florida peninsula along the coastlines of Alabama, Mississippi, Louisiana, Texas and Mexico, down to the base of the Yucatan peninsula, nearly three-quarters of a perfect circle can be seen. The circle has only been marred by the intrusion of deltaic sediments brought in by rivers, and by the fact that a second basin rim, which forms Cuba and the Yucatan peninsula, exploded partly into the Gulf basin.

Another nearby example is the western half of the United States. It has long been evident to geologists that the Great Salt Lake represents the center of a large basin. But probably few have noticed that if dividers are placed with one point in the Great Salt Lake, the other point will scribe a line that neatly parallels the entire west coast of the United States. But when you think about it, the center of an explosion basin – being the last place to dry up when the rim is breached – would be the place where vast quantities of salt would be left by the evaporating inland sea. The relationship between the central Great Salt Lake deposit and the rim hundreds of miles away becomes evident.

So it seems fair to say that the greatest catastrophe in the history of the solar system occurred when Aster exploded and bombarded all of her sister planets with great planetoids. Every planet that we have observed still bears the scars of this catastrophe. Although time has softened the shapes of these scars on Earth, they still remain. Aster's planetoids gave Earth the rough form of the continents that still exist today.

Aster set the stage upon which the drama of life as we know it was to be played.

5

Vulcanism

Ancient Catastrophic Vulcanism

The latter part of the eighteenth century saw a tremendous volcanic catastrophe in Iceland. On June 11, 1783, after a series of violent earthquakes near Mt Skapta, an immense outpouring of lava began along a 10-mile line, later named the Laki Fissure. Lava filled and overflowed the Skapta River Gorge, which was 600 feet deep and 200 feet wide in places. The lava flow was followed by another a week later, and a third on August 3. So great was their volume that they spread out in great floods 15 miles wide and 100 feet deep.

This is the only instance within historical time of a fissure eruption or lava flood. The United States has had many ancient lava floods, however. In the Pacific Northwest, a tremendous flow, covering most of Washington, Oregon and large sections of California and Idaho, is now called the Columbia Plateau. In some sections, more than a mile thickness of rock was ultimately built up. In the canyon of the Snake River, Idaho, granite hills from 2000 to 2500 feet high have been covered by 1500 feet of basalt from these flows. The great Palisades – overlooking the Hudson River opposite New York City – is merely the thin edge of such an ancient lava flood.

One hundred years after the lava flood in Iceland, another terrible volcanic catastrophe took place. The island of Krakatoa, in the Sunda Strait between Java and Sumatra, was completely destroyed by a series of volcanic explosions. One explosion blew a column of volcanic debris 17 miles up into the sky. The dust rose to such heights it was spread around the Earth and took more than two years to fall. During those two years, sunsets were abnormally colored all over the world. The sound of the explosion was detected 3000 miles away. The year following the eruption of Krakatoa was known as the 'year without a summer', because the dust in the air so shielded the Earth from the Sun's rays that temperatures were abnormally low; and some farming areas had no growing seasons at all.

The eruption of Krakatoa left behind a wide, shallow depression or 'caldera'. Ordinary volcanoes vent their materials and leave a cone-shaped mound surrounding a 'crater'. A crater is a steep-walled depression with a floor seldom over 1000 feet in diameter, and a depth seldom more than 300 feet. A caldera is also a steep-walled depression, usually circular, but with a diameter many times greater. Most calderas, in fact, are more than a mile in diameter; hundreds are eight or more miles across and thousands of feet deep.

Many readers will recall the TV documentaries about the isolated herds of rhinoceroses. These were filmed in the Ngorongoro caldera because its entire population of life forms has been isolated but able to live for many generations in this enormous steep-walled basin.

There is a great void between the maximum diameters of volcanic craters – about 1000 feet – and the minimum diameters of volcanic calderas – about a mile. If volcanic activity in the past simply ranged between small and large eruptions, we would expect to see a full range of crater and caldera sizes smoothly distributed from the smallest up to the largest. Instead, we find two distinctly different groups of shapes and sizes composing the record of past vulcanism. It is reasonable to wonder if these are the products of two entirely different types of vulcanism.

The explosion of Krakatoa is the only modern example in existence of the formation of a caldera. There are uncounted thousands of other calderas on the Earth's surface, but these were all made before the dawn of recorded history. Some of those calderas indicate eruptions many times more powerful than that of Krakatoa.

Measurements made of the caldera left by Krakatoa indicate that, of the original island, about five cubic miles of material had been blown so far away that it could not be found and identified. At Crater Lake caldera in Oregon, the volume of missing material is about 10 cubic miles. At Valles caldera in New Mexico, a volume is missing of about 50 cubic miles. The Toba Depression in Sumatra is a giant caldera which, when it erupted, blew 500 cubic miles of material completely out of the region. That much dust in the atmosphere could have darkened the skies for centuries.

The explosion of Krakatoa and the 1783 lava flood in Iceland are therefore apparently only faint echoes of a type of ancient volcanic activity that has left its scars all over the world. The volcanic eruptions that are common today bear little resemblance to the volcanic activity that the ancient world experienced. Although volcanoes are still locally destructive, the giant explosions and lava floods that characterized ancient vulcanism are extremely rare. What could have caused these ancient catastrophic eruptions?

Earth's Internal Heat

Ever since the birth of the science of geology, we have attempted to learn the structure of the Earth. Despite great advances in technology, conjectures on the nature of the Earth's interior are still highly speculative, because they are derived from secondhand sources. The radius of the Earth is about 4000 miles, but we have only been able to drill down about eight miles below its surface. The structure of the Earth below that level can only be inferred from surface phenomena or the actions of Earth as a whole.

Early geologists who observed volcanoes conceived of a molten Earth overlaid with a thin crust of cool material upon which we live. Religious training that Hell was a hot place, deep in the bowels of the Earth, may have predisposed them to this conclusion. Scientific proposals of the formation of the Earth developed, suggesting that the Earth condensed from hot gases, possibly thrown off by the Sun,

This was in agreement with the idea that the interior of the Earth was hot and molten. But all concepts of an Earth that is gradually cooling down from its original heat of formation must collapse in the face of some simple facts.

There is continuing vulcanism. It bursts through the Earth's crust with great violence, builds cones or islands or explosions, then diminishes, only to burst out again another day. Since the materials that are evidence of the sources of vulcanism's power (magma, lavas, ash and gases) all have in common the physical fact that they are at their lowest enthalpy in the solid state, they should be shrinking and causing great holes to open downward into the Earth's crust, instead of the opposite, if the Earth were cooling.

Obviously then, Earth is at the very least continuing to receive or generate some kind of internal heat that sustains its continuing vulcanism.

The concept of the formation of the solar system that currently has the broadest support is the dust-cloud hypothesis, discussed earlier. This hypothesis suggests that the Earth and the other planets were formed from cold dusts, compacted by gravity and sorted by electromagnetism. Despite the fact that geology textbooks espouse this idea, however, they still continue to support other hypotheses that are based on an Earth with a molten interior. How the Earth changed from a cold ball of dust to a hot ball of molten material with a thin, cool crust is barely touched on.

The velocity of sound waves passing through the interior of the Earth has been used to infer its structure. Geophysicists have studied the data to determine whether the interior of the Earth is solid, crushed solid or liquid. Apparently they have never considered the possibility that the outer core of the Earth is made up of a compacted cold slurry, and that a slurry might transmit sound waves in much the same manner as would molten rock.

In fact, another model of the interior of the Earth can be supported by the sound-wave data. This model would include a crust of crushed rock; a mantle of solidified rock with isolated hot spots; an outer core of cold slurry; and an inner core of solid or densely compacted nickel-iron that is also cold.

In a chapter on mountain building, this book will present further evidence to show that the Earth is essentially solid rather than liquid. At the moment, however, it is important to realize that the heat that powers volcanic activity need not be derived from deep within the Earth. It is quite possible that all of the Earth's heat is generated within 50 miles of its surface, and that the interior is cool. But how, then, is that heat generated?

It is axiomatic in physics that energy can neither be created nor destroyed. It may be changed in form, as a generator transforms mechanical energy into electrical energy plus heat, or as an electric motor does the reverse, but despite these transformations, the net amount of energy involved never changes. The Earth, because of its great mass and the speed at which it rotates on its axis and revolves around the Sun, has a lot of energy stored up. Nearly all the energy that first started the motion of the Earth is still in the Earth, stored in the form of momentum. If the rotation of the Earth on its axis were to slow down from its present rate of about 1000 miles per hour at the Equator, or if the speed with which the Earth revolves around the Sun were slowed down from its present rate of 18 miles per second, the energy from the momentum would have to change to some other form, probably heat.

In fact, the rotation of the Earth is slowing down, and our days become minutely longer all the time. Some propose that the day was once about 21 hours long, back in the Cretaceous Period, and that slowing of the Earth's rotation has lengthened the day to its present 24-hour period. It is believed that only part of this slowing of the Earth's period of rotation is due to some of the Earth's momentum being used up by tidal friction.

The magnetic north pole does not coincide with the axis of rotation of the Earth. To a navigator, this means that the reading of the magnetic compass must be corrected in most of the world in order to know where true north is. It also means that as the Earth travels through the solar wind, its magnetic field appears to be constantly oscillating. This oscillating magnetic field creates a kind of electromagnetic 'friction' between the solar wind and the electrical currents that regularly flow through the Earth. This 'friction', called 'eddy currents', builds up tremendous heat on the surface of conductive materials in general, and the Earth appears to be no exception. Through the eddy currents, the Earth's momentum is changed into heat, and this change from momentum to heat causes the Earth to slow down.

In May, 1973, the British science magazine *Nature* published a report that on August 4, 1972, the greatest solar storm ever recorded occurred on the surface of the Sun. Four days later, the instruments of the U.S. Naval Observatory recorded that the day had lengthened 10 times more than would have normally been expected. Apparently the currents sent out by the solar storm caused a tremendous increase in the eddy currents in the Earth. This increase in eddy currents sapped the momentum of the Earth and slowed its rotation by a perceptible amount that must have generated an equivalent amount of internal heat.

Perhaps, too, some heat is produced by atomic energy. Some textbooks regard it as possible that radioactive decay accounts for much of the heat being produced under the Earth's surface. This hypothesis is given weight by the proposal of Francis Perrin, former commissioner of the French Atomic Energy Commission, that evidence of a prehistoric spontaneous nuclear reaction (in the Oklo mine in Gabon, Africa) was discovered by researchers at the French uranium-enrichment center. This chain reaction was near the surface but did not cause an atomic explosion, apparently because of the moderating effect of water in the soil.

Even nuclear fusion, which the Sun and the hydrogen bomb use as a source of power, may be possible at the high pressures that are present under the surface of the Earth.

What happens to the heat that eddy currents and nuclear activity generate in the Earth's mantle? The answer is difficult to determine. Undoubtedly, some of it is conducted to the surface and radiated into space. But perhaps some of it remains in the mantle, building up over long periods of time, unable to be dissipated because of the Earth's insulating qualities. This heat, combined with the heat of radioactive decay, may build up tremendous pressure within the Earth. The great heat and pressure may initiate radioactive fission and fusion, resulting in huge volcanic eruptions – explosions that blow powdered rock, metals and radioactive materials high into the skies, as tremendous floods of lava cover the land and the sea floors. When the Earth has released its internal pressures, volcanic activity then changes to the comparatively peaceful form that it takes today, and only the scars – the calderas and lava fields – are left. It is probably fortunate for Earth that it has been able to release these internal pressures periodically. Devastating as they are to life forms on the surface, the alternatives, such as Aster's complete destruction, make these blow-offs a form of 'safety valve'.

The Origin of Mineral Deposits

The calderas and lava floods that dot the Earth indicate that volcanic activity was once very different from that seen today. Perhaps that volcanic activity can account for the unusual mineral deposits that are found all over the world.

It is incorrect to think that mineral deposits were formed at the same time as the Earth itself. In fact, deposits have always been formed later. There is a mystery about placer iron ore because of the great areas over which it is deposited. It is not possible to attribute mechanisms such as precipitation from water to deposits of such great size. The Big Seam of Clinton, near Birmingham, Alabama, is an iron ore deposit over 30 feet thick in some places, having an average thickness of 10 feet. This deposit alone is 10 miles wide and 50 miles long, in a region where outcrops are almost continuous for some 700 miles. The total weight of the single deposit is about five billion tons. The Wanaba Basin in Newfoundland holds about seven billion tons of iron ore, and the Lorraine area of France has over five billion tons. Placer iron accumulation on such a vast scale simply cannot be explained by microorganisms in water causing the iron to be extracted.

Then what were the conditions under which these placer deposits were laid down? Where did these vast amounts of nearly pure iron come from? The deposition of other metal placer ores such as copper and aluminum, and of many clays, is equally surprising because of the purity of the deposits. If these deposits were laid down over vast periods of time, it is reasonable to conclude that they would be decomposed and contaminated. Since they are relatively free of stray materials, they must have been concentrated quickly. But how?

Is it possible that a volcano can act as a huge fractionating tower – in effect, a natural smelter? Under the conditions of extreme vulcanism which the Earth had endured in the past, perhaps a volcano could, for a time, pour out pure metals or minerals in the form of dusts. Perhaps on occasion, iron dust was blasted out of ancient calderas to fall in pure deposits as we find it today.

Many of the minerals that are found in the world today are supposedly the products of the decomposition of other minerals. For instance, the clays illite, kaolinite and montmorillonite and the potassic mica muscovite are supposedly derived from the decomposition of igneous rocks such as granite or basalt. These minerals, however, are valuable because they can be found in placer deposits that are remarkably pure. In the time it would take granite or basalt to decompose in sufficient quantities to make sizable deposits of these materials, many impurities could also find their way into the deposits. Despite this, deposits are found in nature relatively free of the contamination of other minerals or organic matter.

In addition to this, uniformitarians are not clear as to exactly how this decomposition takes place. They propose that if granite is exposed to weathering, feldspars are the first minerals to be broken down. But mineralogists and soil scientists still cannot explain the process by which the feldspars break down. They suggest that, theoretically, water plus carbon dioxide yields carbonic acid, which in turn yields hydrogen plus a bicarbonate ion; that orthoclase then comes in contact with the hydrogen ions; and that orthoclase plus hydrogen ions plus water yields clay.

Although the chemicals amd minerals mentioned in the preceding are neither exotic nor rare, the reaction which supposedly takes place among them is unlikely. Chemists advise that these reactions would be extremely difficult to bring about, even under the best of laboratory conditions. It is simply not conceivable that they would frequently occur under the conditions that nature offers at present.

If the clays and ores are not produced by decomposition, then where did they come from? And why are they found in such a pure state? The answer may lie in the combined actions of volcanoes and ice sheets.

Perhaps, during ice ages, volcanic smelters briefly rained down pure dusts of these minerals on the surfaces of the great continental ice sheets. The periodic melting of the surface of these ice sheets could then carry the collection of minerals away with their meltwaters, thus concentrating the pure ores. This speculation has the virtue that the process is easily duplicated in miniature by human technicians – and who should say that nature may not elect the most direct processes of ore refinement?

The ocean floor, too, has its share of mysteriously deposited minerals. The presence of nickel in the red clay of the Pacific is puzzling, as are the nodules of manganese in some places. The nodules, which also contain nickel and cobalt, are up to three feet in diameter. They have been found as thick coatings on a large proportion of the rocks dredged from both the Atlantic and the Pacific. It is rare that any manganese is found in cores from the ocean bottom, except at the top of the cores.

Perhaps the manganese nodules, like the clays and the great iron deposits, represent concentrations of metal dusts in the oceans and are the products of vulcanism on a scale that no modern human has ever seen. Perhaps the Earth's mantle, overheated by eddy currents and by the decay of radioactive materials, built up pressures beyond the limit of the weight of the crust. Lava floods poured across thousands of square miles, burning and burying everything in their paths. Smelter-like caldera jets spewed geysers of pure minerals into the air, laying down placer deposits of remarkably pure ores. Old stories of people gathering diamonds, gold, copper and other metals from the surface of the land would explain why these bonanzas today may now be found on the surface only of the inaccessible ocean floors.

The catastrophic vulcanism that spawned these actions may have come close to destroying the ancient world. But it began a series of steps that would prepare the surface for its next inhabitants.

6

Changing Sea Levels

Under the Oceans

Georges Cuvier, the great French paleontologist, observed that frequently continents which have been 'dry have been again covered by the waters, in consequence either of their being engulfed in the abyss, or the sea merely having risen over them'. His studies led him to believe that these great inundations and emergences were not limited to the time since life appeared on Earth, but had occurred throughout time.

Because he was a careful and precise scientist, Cuvier equally emphasized the two possible explanations for continents frequently being inundated and emerging from the sea: that the land may have elevated and fallen; or that the sea levels rose and fell. There are many conditions that can cause localized elevation or subsidence of land, but there is convincing evidence that only vast changes in the levels of the oceans of the world can be the cause of the continental submergences.

Glomar Challenger, a ship designed to drill material from beneath the deepest oceans, brought up cores from the floors of the Mediterranean Sea, the Atlantic and Pacific Oceans and the Gulf of Mexico – with some startling results. Many have read the *Reader's Digest* article about drilling the two-miles-deep floor of the Mediterranean where they found that the oldest sediment at the bottom of the cores proved that the Mediterranean had come into existence as an arm of the Indian Ocean, since it contained organisms that are found nowhere else. But above this primordial ooze they were startled to find beds of salt and anhydrite one mile thick. The finding of these materials – often left as a residue when seas dry up and therefore named 'evaporites' – prompted the instant hypothesis that the Mediterranean had periodically dried up completely.

Similarly drillings in the Gulf of Mexico have established that its floor is also underlaid by vast beds of salt and anhydrites, implying that this body of water, too, has been completed isolated from the rest of the oceans of the world at various times.

Topographical maps show clearly that the Mediterranean Sea's original outlet to the Indian Ocean through the Red Sea was gradually cut off by the growth of the delta of the Nile River. The Mediterranean subsequently developed a passage to the Atlantic Ocean through the Straits of Gibraltar.

The Straits will carry water to and from the Ocean, preventing the formation of salt beds, until the Atlantic falls more than 650 feet below its present level. Textbook estimates of the fall of ocean levels caused by the growth of continental ice sheets are on the order of 600 feet. The closeness of these two figures makes it easy to believe that the formation of continental ice sheets only in arctic and temperate regions is sufficient to cut off the Mediterranean from its oceanic outlet.

The evidence of the Gulf of Mexico, however, forces a major modification of this idea. In order for that body of water to be cut off from the Atlantic Ocean, a fall in sea level of nearly 6000 feet is required. Only when the Atlantic falls that far is the Gulf isolated from it, theoretically allowing the evaporites to accumulate.

Columbia University published many surprising results of their oceanic explorations. Among those findings was the location of prehistoric beach sand on the slopes of the deeply submerged Mid-Atlantic Ridge. 'Beach sand' is a term for sand produced by the action of a surf line or river flow, turning rough rocks and crystals into well-rounded pebbles and sand. This beach sand is usually found to be stratified, or sorted into layers of different coarseness. Yet the sand which was found was in one spot two miles under water and, in another, under nearly three and one-half miles of water – and hundreds of miles from the nearest land. It has since been found that sand beds, some of them as clean as those found on the beaches and in the shallow water near shore, were well represented on portions of the deep-sea floor. The surprise was even greater when it was discovered that these sand layers contain shells of foraminifera that live only on the bottom of seas no more than a few hundred feet deep. Piston coring devices obtained long cores which revealed that the sand layers alternate in many places with typical deep-sea deposits.

In addition, mudstone has been brought up from a depth of 3600 feet. Mudstone is a weak, loosely consolidated rock which is deposited in no more than 600 foot deep marine environments. Furthermore, it is too fragile to have survived transportation to that spot in deep water.

At that same depth, the explorers found granite and other sedimentary rocks that are usually associated with the continents. These rocks were in most cases deeply scratched and striated, as if they had been caught up in a glacier. Yet they were found far from land, and deep under water.

Estimates of the thickness of sediments that mantle the deep-ocean floors were made, based on the amount of sediment now contributed annually to the ocean and on the estimates of the age of the oceans. From these figures a marine geologist computed that there should be an average thickness of about two miles of sediment. However, when seagoing geophysicists started using the technique of the oil geologists to send sound waves through the ocean-floor sediment, it was found that the thickness was far less than supposed. In fact, in some places there was virtually no sediment at all.

Instantly, the uniformitarians seized upon this finding to assume that the rock which underlies the ocean-bottom sediments was basalt. Then they proposed that the thinness of the sediments was another proof that the basaltic 'conveyor belt' of the spreading sea floors hypothesis was carrying the basalt under the continents too fast for it to acquire a thick cover of sediment.

But when drilling expeditions penetrated the rock of the ocean floors, that hypothesis was destroyed. For example, in the Pacific Ocean southeast of Japan, while drilling the ocean floor beneath 20,000 feet of water, researchers found that the rock of the Marianas Basin was not basalt but *limestone*! Limestone, made of the skeletons of tiny marine animals, forms only in shallow seas, no more than a few hundred feet deep. This is because the animals only live near the surface, and when they die in very deep waters, their shells dissolve as they sink before they can reach the bottom to form limestone. Yet here lay beds of limestone hundreds of feet thick, about four miles below the ocean surface. Beneath the limestone was another layer of abyssal ooze similar to that which covered it. Sonar exploration has indicated that beneath these beds lie still other layers of limestone, separated by more layers of deep-sea ooze. That sequence is identical to the alternating shallow and deep beds found in the depths of the Atlantic.

Widespread ocean-floor probing has now established that limestone in deep oceanic basins is the rule rather than the exception. The hypothetical basalt has lost its significance as proof of continental drift.

Drilling of the sea floors in many places has also shown that some layers of sediments which should be present are missing completely. Uniformitarians have proposed that a great rush of cold arctic water from melting ice sheets killed the life forms that would have made up the missing sediment layers. These reports of sea-floor peculiarities were but the first of what have become almost routine new disclosures as drilling and exploration probe the oceanic deeps. East of St Augustine, Florida, for example, from under 9000 feet of water *Glomar Challenger* brought up reef cores containing evidence of having been above sea level – not once but many times.

The findings of sand, striated rocks, mudstones, alternating shallow-deep layers and evidence of extreme cold on the deep ocean floor present an inescapable conclusion. Either the ocean basins must have periodically risen three miles; or the sea level must have sometimes been three miles lower than now. And if the oceans were sometimes three miles lower, where could all that water have gone?

The uniformitarian approach to estimating the extent of ice sheets, and therefore the fall of sea level, calls to mind a story of Thomas Edison, a self-taught genius. When he asked one of his engineers to determine the volume of a light-bulb blank, the man tackled the problem with careful measurements and calculus. Before he had gotten very far, Edison took the bulb from him and filled it with water. He then poured the water from the bulb into a graduated beaker and read off the volume. The engineer quickly learned the value of a direct and simple approach.

In a similar manner, the extent of ice on land can be determined from the fall in sea level much more easily than by trying to estimate which land debris is glacial – and how thick the ice may have been.

Uniformitarians use the ice area and thickness technique to estimate that sea levels fell no more than 600 feet during the greatest ice ages. But evidence on the ocean floors indicates that sea levels have fallen three miles, and several times during the Earth's history. This means that sometimes the ice sheets that stood on the continents were thirty times more massive than was previously suggested. These giant ice sheets must have covered all the present land area of the world, from pole to pole.

These were truly the 'ice ages'.

Submarine Canyons

Still more evidence that the sea level was once three miles lower than it is today may be found in the form of the undersea canyons. Canyons on land are formed by water running downhill at high speed. The materials which the running water carries erode deep V-shaped canyons, even in solid rock. A few canyons on land are carved by glaciers into a characteristic U-shape, but the difference between these and the canyons cut by running water is unmistakable. Also, when a river reaches a body of still water, it ceases motion and releases the mud and sand that the running water held. This material piles up to become a delta.

Everyone agrees that V-shaped canyons on land are caused by running water. But there are other V-shaped canyons that lie deep under the oceans where, according to uniformitarians, rivers could never flow because the floors have always been covered by water. In fact, all the continental areas of the world are surrounded by these canyons, in great numbers, all leading to the deepest parts of the ocean.

Two widely separated examples of undersea canyons are those leading from the mouths of the Congo River in Africa and from Tokyo Bay. The Congo canyon is continuous all along the length of the river out into the ocean, and down the continental slopes to a depth of three miles. Dr Francis P. Shepard of the Scripps Oceanographic Institute studied the lower end of that canyon and found a great river delta – complete with many distributaries. The finding is consonant with that of an ordinary river delta, except that this one lies miles under the oceans rather than at its surface. Dr Shepard further noted that echo-sounding equipment recently put into widespread use shows that some of the canyons have walls that are even deeper than those of the Grand Canyon of the Colorado. And the grand-daddy of them all may be the greatest drowned river system of the world. This system, dominated by the 'Mid-Ocean Canyon', runs along the floor of the Atlantic all the way from both sides of Greenland, to the Sargasso Basin, east of Florida.

This drowned river system is 2500 miles long and includes many impressive tributaries. The major tributary of the system was probably the Hudson River and its related Submarine Canyon of the same name. It is one of the most famous of the undersea canyons, visible as a continuous river bed on marine charts from near Albany, New York, past New York City and around Sandy Hook, New Jersey. From here it runs a thousand miles to the southeast to join the Mid-Ocean Canyon somewhere along the way to the Sargasso Basin. Fishermen working the Jersey shore have long cursed this canyon because its deep chasm suddenly interrupts the flat bottom of the fishing grounds, and unwary captains have seen nets, trawls and fish drop away to their great loss.

Other coastal canyons also were North American tributaries – apparently draining the great ice sheets. Meltwaters from the southern part of the United States poured into the Gulf of Mexico, then east, around the peninsula of Florida, across the Bahama Island shallows and down into the Sargasso. Along the way, these waters carved their most colossal works. Around the Bahama Islands are found submarine canyons more spectacular than the Hudson. Here tremendous flows of water created waterfalls in much the same manner as Niagara – but on a scale that dwarfs Niagara by comparison. These canyons, named Exuma Sound and the Tongue of the Ocean, are now completely covered with water. Passengers sailing the clear Bahamaian waters can see the ocean bottom until they reach these canyons, where the water depth instantly changes from 30 feet to about 6000 feet.

When the ice sheets were melting, these sudden drops were titanic waterfalls. The Tongue of the Ocean begins as a bowl 45 miles in diameter. Around the edge of the bowl, water fell a mile in an unbroken cascade – and then roared down a canyon 150 miles long with mile-high walls on either side. By comparison, the waters of mighty Niagara fall only 175 feet, less than one-thirtieth of the ancient height of the Tongue of the Ocean.

Turbidites and Platforms

Uniformitarians, who do not allow that sea levels were ever substantially lower than they are now, explain the submarine canyons by ascribing their erosion to 'turbidity currents'. Turbidity currents are undersea mudslides whose existence was originally postulated to explain turbidites – the well-sorted sedimentary deposits that cover great areas of the sea floors. Supposedly, these mudslides stirred up clouds of sediments. The largest grains in the cloud, the gravels and sands, settled first. These were covered by successively smaller particles that settled more slowly. However, turbidites are found in many areas – not just at the lower reaches of steep slopes – so a better explanation of their origin is needed.

The claim that fast mudslides, slipping down the gentle slopes of the sea floor can erode rock, just as a river does, is unreasonable. On land, mud can move quickly only down fairly steep slopes, and then only for short distances. In addition, mud tends to fill in rather than erode the land over which it flows. The hundreds of miles of unsloping floors of undersea canyons and the vast undersea areas all over the world that look like wide, flat river deltas, far exceed the distance such mudslides could reasonably be expected to travel. On a deeply submerged submarine delta off Monterey Bay there has been found a meander. A meander is an S-shaped river bed carved by slowly moving water flowing over nearly level ground. A high-speed mudslide could not produce such a winding riverbed.

Probably turbidites are caused by two low-sea-level phenomena: the sorting action of rivers flowing in the now-submarine canyons and the result of a moving surf line. As the changing sea level causes a surf line to move uphill from a given area, the sediments that the surf line has churned up settle in the still waters behind it. The particles settle in diminishing order of size, just as in turbidites.

The churning action of a moving surf line not only deposits turbidites, but also tends to gentle the slopes during its traverses. On a small scale this may be seen at the beach. The waves erode the high spots and fill the low, so that a gentle slope results. The products of this erosion are often carried far out beyond the breakers, so that the slope is much wider than can be seen above water.

During the eons of ice-age-caused sea level changes, and on a very large scale, these moving surf lines have produced the combination of coastal plains plus continental shelves which presently surround the continents of the world. Other surf lines have also acted on the interiors of the continents during periods when they are flooded by continental seas, which produced prairies, steppes and plains. Together, these coastal and inland areas combine to form an erosion 'platform' of gently sloping land that surrounds all the steeper highlands of the world.

These platforms on today's continents are the lands on which we live and fish, and collectively they account for about 30 percent of the Earth's surface area. In effect, they are a border between all the mountains of the world and the deep water of the oceans.

A trip along a line drawn from a high coastal mountain to a spot where the ocean is five miles deep shows the nature of the platform and also provides a surprise. First, there is a steep drop from the peak of the mountain to the platform. Then the gentle slope of the prairies and coastal plains, the beachline and the submerged continental shelf. At the end of this gentle descent, there is a comparatively steep grade: the continental slope. This steeper grade continues to a depth of approximately three miles, where *the grade once again becomes gentle*! This gentle grade resembles the platform that lies above it, and is called the 'continental rise'. In fact, it is another platform, but one that is closer to the deeper ocean basins. This lower platform continues for some distance, after which the ocean floor again drops off to its deepest regions.

Like the upper platform, the lower platform rings the oceans – but at a depth of about three miles. It is not exclusively related to the continents, however, because the lower platform surrounds all islands and submerged ridges, as well as the continents, at that depth.

Surf erosion of highlands and the deposition of sediments have given the upper platform about 30 percent of the Earth's surface area. The lower platform has about 23 percent of the Earth's surface. Probably the lower platform is just what it looks like. It is the work of many ice-age surf lines, fluctuating about three miles below the present sea level. This explanation of the formation of the lower platform solves many of the mysteries raised by the other findings on the sea floors.

It is from approximately this depth that beach sand is often found. If the lower platform is a former shoreline, beach sand is exactly what one would expect to find at this depth. Then there are those deep-sea corings that revealed layers of limestone or clean beach sand alternating with abyssal deposits. Since it has been shown that the Earth has sustained at least twenty ice ages, the alternation of deep-water deposits with shallow deposits loses its mystery. The alternation merely reflects the change from high to low sea levels as the ice sheets emptied or refilled the oceanic basins. The cycle from high to low sea levels also accounts for the missing sediment layer. Changing an area from deep ocean floor to shallow sea or even dry land would certainly cause a drastic modification of the fossil deposits.

But more than any other fact, the undersea river canyons confirm that the lower platform marks the lower extent of the seas. It is to this level that the canyons run. From the continental mountains they are continuous across the upper platform and down the slopes to the lower platform – but no farther. Here thos ancient rivers met the still waters of the shrunken seas and deposited their sediments in deltas that still lie undisturbed on the sea floor.

Unique evidence of changing sea levels may be dredged from submerged seamounts or guyots. These flat-topped mountains are apparently formed when volcanic activity on the ocean floor builds a great cone of cinders and lava. If the cone extends above the present-day surface, it may become useful for habitation, such as the Hawaiian Islands, which are very large volcanic cones. But many of these cones have stopped short of attaining the area and mass that projects them above present sea levels. Those that have been very recently formed retain the shape of volcanic cones on land, coming up to nearly a point, in which the crater is located.

But there are thousands of more ancient cones that have not retained their pointed tops. These are the ones that are named guyots to denote their planed-off tops and to honor an early oceanic scientist of that name. In the past, these planed-off cinder cones have proved to be a mystery to those who wondered at the great variety of levels to which they were flattened. Not only are they cut off at different altitudes, many are flattened far below the minus-600-feet that uniformitarians allowed the sea levels to fall. So a great number of hypotheses of their origin were forced to evolve. Some had them growing above present sea level, then being planed flat by the surf, then sinking hundreds of feet below the minus-600-foot line. But they didn't sink the same distance, so that explanation was unsatisfactory.

It should now be possible to see that these cones are the direct proof of both the fact of great sea level changes and the time when the volcano ceased its growth. It is as originally proposed. The cones grew until they penetrated the sea level of that time. As long as their tips remained above sea level, they were subjected to the erosive forces of weather and waves. But when the erosion – or changing sea levels – submerged them below the planing action, they ceased to be flattened and could consolidate. Consolidation could be either by the change of lava from an easily eroded glass to a crystallized rock or by the cementing effect that calcareous microorganisms have when they form limestone.

It will now be possible, therefore, to study the types of organisms that first consolidated the flat surface of these guyots and to learn from them when the flattened surface was first submerged. Then this information may be correlated with the height of each guyot. In view of the thousands of submerged table-mounts, from this study a very accurate charting should be derived of the recent sea level changes – if not some of the earlier ones.

In 1975, the results of the Franco-American Mid-Ocean Undersea Study (FAMOUS) began to be published. This survey of the miles-deep Mid-Atlantic Ridge southwest of the Azores set out to prove – by direct observation from deep-diving submarines – that the supposed line of separation between diverging 'plates' of the plate tectonics (drifting continents) hypothesis did, in fact, exist.

It was not surprising, therefore, that so single-minded a purpose resulted in the drafting of articles and news releases confirming their success at so doing. But the celebration was very subdued, and all of the published data and photographs showed only an underwater scene typical of volcanic mountain peaks and valleys in which recent submarine eruptions had formed 'pillow' lava tubes peculiar to shallow underwater lava flows.

One piece of unheralded data they returned with was a detailed topographic chart of an area about 20 miles by 20 miles on a section of the Ridge. Analysis of this chart does not confirm the predicted topography of the Drifters – in which there is proposed accordion-pleated folds (the 'rift' valleys) on each side of the Ridge, sharply incised by cross-fractures (the transverse faults) that were expected to show the process of plate growth. Instead, the contours reveal a 'typical' mountain valley, running generally north–south for a distance of about 15 miles. In fact, the valley is so 'typical' of those above sea level that it is readily apparent that if the ocean were to be removed, this valley would contain a mountain lake at an altitude that is now minus 2700 meters, and that the overflow from that lake would spill off through a mountain pass to the east of the lake's northern portion.

It is fascinating, therefore, to trace the course that the lake's overflow would necessarily take over this terrain and to find that a river canyon has in fact been cut through this mountain pass, just as such canyons are typically formed by mountain lake runoffs everywhere above sea level.

Sea-floor Migrations

The hypothesis of plate tectonics or drifting continents holds that the continents float on the liquid-rock mantle of the Earth like wood on water. Adherents of this idea also believe that the continents can drift about over millions of years, and that all of them were once part of a single continuous landmass that was torn apart.

Devonian life forms found in the Old Red Sandstone of Europe are identical to those found in a similar sandstone in the Catskill Mountains of the United States. Proponents of the drifting continents hypothesis have published many charts which extrapolate this finding into the idea that the ancient Appalachian Mountains had their western slopes in North America and their eastern slopes in Great Britain and Scandinavia. For them, this conclusion is unavoidable because during Devonian times these life forms were not spread all the way across North America and Eurasia. Therefore, the Bering Sea land bridge, which has been proposed as a migration route for later life forms, cannot explain this particular arrangement of fossils.

Other types of fossil evidence have been presented to support the same hypothesis. *Glossopteris*, named for the tonguelike leaves of the seed fern, has been found in South America, South Africa, Australia, India and within 300 miles of the South Pole in Antarctica. *Mesosaurus*, a toothed early reptile of the late Permian Period, lived in the water and left its remains in Brazil and South Africa. Although it was aquatic, most paleontologists do not believe that it could have made the trip across the South Atlantic. *Lystrosaurus* was nonmarine, yet its remains are reported from South Africa and from Antarctica's Alesandra Range. The similarity of fossils in these places in the southern hemisphere that are now remote from each other suggests to uniformitarians that these places were at one time a single, continuous land mass, now drifted apart. Since rock formations also are to a great extent characterized by the fossils within them, this similarity has led some to the conclusion that they are derived from the same source.

On the Indian peninsula lies a sequence of rocks known as the Gondwana system. Beds of similar nature and age are found in South Africa, Malagasy, South America, the Falkland Islands, Australia and Antarctica. Some of these similarities are so striking that uniformitarians accept only one interpretation: the various southern lands must once have been part of a single landmass, a great southern continent called Gondwanaland.

Uniformitarians hold that after the continents were separated, migrations between the Old and New Worlds were accomplished across the Aleutian Islands or the Bering Straits, between Siberia and Alaska. This suggestion has certain serious drawbacks, however.

On a world map, the distance between the Aleutian Islands seems quite short, making the possibility of life forms hopping from island to island from Asia to North America a plausible idea. On the ground, however, the distance between the islands is many hundreds of miles, and the water in between is so cold that only a few mammals can survive its temperatures for more than mere minutes. The idea that mammoths, horses, camels and naked humans could survive this crossing ascribes to them supernatural endurance.

So, the uniformitarians look farther north to the Bering Straits, where water is today only 600 feet deep. Presumably, the often-proposed ice age 600-foot drop in sea level would make this passage dry and crossable. Dry it would be, but crossable it would not be. Even a minor ice sheet increase that lowered sea levels only 600 feet would cause that area to be buried by the expansion of the Arctic Polar Ice Cap. A few stray animals might scramble across the multi-thousand-mile, barren, icy bridge, but the mass migrations that populated the plains of Europe and North America with identical animals would be impossible under these circumstances. Clearly we must look elsewhere for an explanation of the migrations between the Old and New Worlds of the Northern Hemisphere.

Some who reject the idea of drifting continents still believe that there was, early in the history of the Earth, a connection among the continents of the southern half of the world. They have proposed that Gondwanaland was connected by narrow bridges of land that have since sunk beneath the oceans. That hypothesis does contain one valuable insight. It is not really necessary to have had the continents abut each other to explain the life form similarities among them. It is only necessary that there be migration routes between them by land bridges. Land bridges could account for the worldwide distribution of the *Glossopteris* flora, as well as for the presence of *Mesosaurus* and *Lystrosaurus*, on continents which now have no connection whatsoever. Land bridges could also explain the similarities of the fossils in the Gondwana system in different parts of the world.

But where did these land bridges come from, and where are they today? They lie deep under the ocean, where they always have been. The three-mile fall of the sea levels of the world which fed the continental ice sheets and isolated the Mediterranean Sea and the Gulf of Mexico also exposed areas of land which usually lie under water. When sea level is lowered this much, South and Central Africa, India, South America and Antarctica are all connected by dry land that was formerly sea bottom.

The three-mile fall in sea level caused most of the sea floors of the world to become lush grasslands and forests. The decomposed sea floor plants and animals, combined with the former abyssal ooze, formed a soil of enormous fertility. The snows that fed the continental ice sheets became rain on the warmer sea floor. This rain, combined with the many rivers of ice sheet runoff that carried water from the continents into the shrunken seas, kept this fertile land well watered and made life abundant.

Many animals, from humans and camels backward into time, were able to walk between Europe and North America across this vast pasturage. Sea-floor charts show that at about the fiftieth parallel an unobstructed path runs across the sea floor between Newfoundland and England when sea level is lowered three miles. This is approximately the route that ocean liners take between these two points. Near the western end of this route, on former plains that are now the Grand Banks fishing area, fishermen have for years netted evidence of this migration in the form of the fossilized bones of prehistoric animals.

During the ice ages, the transatlantic route between England and Labrador may have been too cold for passage by some animals. For these, another path across the Atlantic lay between Dakar in westernmost Africa and Natal in easternmost South America. At those warm latitudes, the path was probably lush savanna or forest all year round.

By either the polar or equatorial routes, plants and animals could range between the icy highlands of Europe and North America, Africa and South America, across the grasslands of the former floor of the Atlantic Ocean. From the South Atlantic Basin they could migrate even farther. By going eastward past the tip of Africa they could enter the basin of the Indian Ocean. The Indian Ocean itself had disappeared, along with the water of other oceans (except for several small seas). South of the Equator, eastward migration to Australia was barred by one of these seas, as can be seen by reference to marine charts today which show the shapes of the ocean floors. But elsewhere animals and plants could range from the Atlantic Basin through all the area of the Indian Ocean Basin.

In the South Pacific Basin, an area of land three times the size of South America was available to the people and animals that were forced here by the ice age. Farther north, these plains continued on the west of the frozen continent of North America. Here, however, the plain was only half the width of its southern portion. There was no block to north–south migration in the Pacific Basin, except for the equatorial heat barrier that apparently stopped some plants and animals such as penguins. However, remnants of the Pacific Ocean made a major barrier to intercontinental migration in an east–west direction. As can be seen by present-day submarine charts, the shrunken Pacific still stretched from the Arctic Polar Ice Cap to that of the Antarctic, hundreds of miles across at the narrowest point. That humans and animals crossed this gap then is as unlikely as their doing so today.

Australia's Isolation

A three-mile fall in sea level not only makes migration to the former sea bottom possible, it makes such migrations absolutely necessary. Such a drop of sea level leaves all the present land masses about three miles above this new sea level. The air on these landmasses is then too thin and cold to permit normal life, and migrations to the former sea floor are necessary for survival.

Most representations of the Earth's surface are greatly exaggerated so that surface features such as mountains and oceans can be discerned. But despite these tall mountains and deep ocean trenches, the Earth is a very smooth sphere. If a polished billiard ball were blown up to the size of the Earth, its surface irregularities would be greater than those of the Earth. Also, Earth's surface is so arranged that 30 percent of the area of the Earth lies within one mile of present sea level. And when sea level falls three miles, 33 percent of the Earth's surface lies within one mile of the new, lower sea level, although it is in a different region. Even though the fall of sea level makes all of the present land areas uninhabitable, enough new land is exposed to permit life to continue on the dried sea bottom.

This vertical change of three miles requires a horizontal movement of thousands of miles, since the slope of the floor of the ocean is so slight. Given that this is the case, it is easy to see why the theory of drifting continents was conceived. Looking at the vast oceans and the similarities of life on widely separated continents, it was correct to assume that some sort of world-wide horizontal displacement of life forms had taken place. It was not the continents that were displaced however, but life itself. Forced by falling temperatures and thinning air to abandon their former homes, all life forms moved thousands of miles across former sea bottoms, to stay at an altitude near sea level where they could survive.

As the peak of the ice age passed, the ice began to melt and the seas began to return to their present levels. The plants and animals again retreated, but not solely along the routes which brought their forebears to the sea bottom. When the seas rose again, *Glossopteris*, *Mesosaurus* and *Lystrosaurus* were thus spread to the widely separated continents on which their fossils are now found. The same sort of migration by other life forms caused the rocks of the Gondwana system to have similar fossils over the southern hemisphere. In a later age, this type of migration would cause a similar dispersion of humans and their cultures in parts of the world now remote from each other.

The lack of placental mammals in Australia has been claimed as evidence that refutes the hypothesis of low sea levels. L. Sprague de Camp noted, in his attack on Atlantis deep-submergence stories, that the distribution of animals in Southeast Asia is 'fatal' to many sunken lands ideas. This is because one of the sharpest animal boundaries on earth is the 'Wallace line' through Indonesia that divides the animals into Oriental and Australian. In Borneo and New Guinea is the Indo-Malayan world of monkeys, cats, buffaloes, elephants and so forth. Across the Wallace line is the realm of the kangaroo – the land of the egg-laying mammals and pouched marsupials.

This difference has been taken to prove that the water barrier of the Celebese, Banda and Timor Seas has been in place since the Mesozoic Era, since any lowering would have let placental mammals into Australia, which did not happen.

If a three-mile fall in sea level made migrations possible among all the other continents, why are the animals of Australia so distinct from those of the rest of the world? Reference to contemporary sea-floor charts show that when sea level falls three miles, the ocean floor between Australia and the more easterly islands of New Zealand, the Fijis, the Gilberts and the Marshalls becomes a vast plain. Animals and plants from all of these highlands can move onto this plain and survive the ice age. But they cannot migrate out of it. To the south, migration is blocked by the expanded Antarctic Polar Ice Cap, which extends into southern Australia. To

How high is the weather zone of the atmosphere? How thick were the ice sheets? How deep are the oceans? They can all be contained within this line.

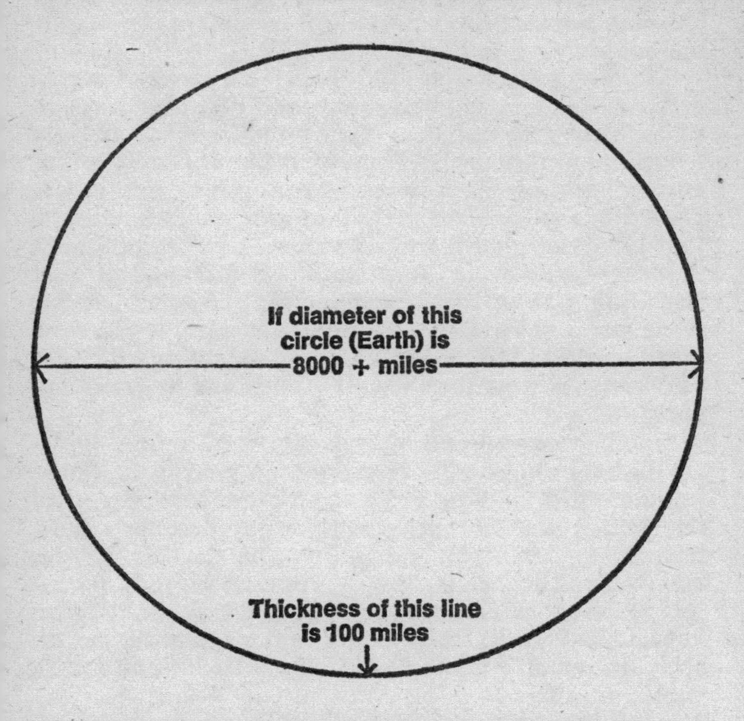

From the top of Mt. Everest to the bottom of the Philippine Deep, the distance is less than 20 miles.

the west, a deep remnant of the Indian Ocean blocks escape, as does the Pacific in the east.

At first glance, one would assume that all this is immaterial, since animals and plants would simply follow Australia's low-sea-level shoreline until they came to the shorelines of Indonesia, etc. But closer inspection of the charts and reflection about ice-age conditions shows why this cannot be.

Across this connecting land there are a series of broad, deep trenches and submarine canyons. These gorges, one of which is the famous Timor Trench that runs for nearly a thousand miles on the sea floor, form an impassable barrier during times of low sea levels, for reasons that will become apparent. The location of this chain of trenches near the Equator means that there is constant runoff from the ice sheets that are covering Australia and Southeast Asia. That runoff turns these trenches and canyons into icy, raging rivers, many miles across. Even during low sea levels, these meltwater-filled trenches thus form an unbroken water barrier that runs from the Indian Ocean to the ice sheet and then to the Pacific.

The Wallace line, when drawn, did not follow any apparent physical boundary. It was merely a line between islands to show a division between distinctive life forms. But when the Wallace line is superimposed on a chart that shows undersea topography, it becomes apparent that the line runs exactly along the line of deep trenches and canyons. Indeed it is possible that these trenches are parts of a drowned river system that was carved by the swift passage of the ice sheet meltwaters hurrying down to the Pacific and Indian Basins.

So, since the Mesozoic Era at least, when sea levels were high, Australian animals were isolated by standing water barriers hundreds of miles wide. And, when the ice sheets lowered the sea levels enough so that animals could cross these straits, they were filled with another water barrier, icy torrents of meltwaters from the frozen highlands.

The logic of Australia's isolation adds its weight to the

synthesis of great sea level changes having been caused by equally massive ice sheets built upon the continents.

Atlantis

We like to believe that somewhere, sometime, there was a land of peace and plenty; a land of beauty and justice; where the cornucopia flowed and the unicorn played. Was this the fabled Atlantis?

Atlantis was first mentioned, according to history as we know it today, by Plato, who recorded that his information came from a manuscript written by Solon, an Athenian lawyer born in 638 B.C. Solon claimed that he was forced into exile from Greece, and that while in Egypt was told of the existence of Atlantis by historians known collectively as the 'wise men of Sais'. According to the custom of the time, Solon was in the process of formalizing this bit of history by writing it in verse when he was struck down by the Grim Reaper.

Plato, recognizing the value of Solon's manuscript, also undertook to complete the record in writing, but he too succumbed before the story could be finished. So we have today only the beginning of these authors' stories, and that portion, devoted as it is to the characters of the period, instead of to defining Atlantis' locale, leaves us with more mystery rather than less.

According to Plato's unfinished *Critias*, The Beginning was marked by the division of Earth among the gods. (The first astronauts, perhaps?) The god Poseidon took possession of a land he named Atlantis. His romance with the mortal woman Cleito resulted in his establishing her upon an island that was surrounded by alternate rings of water and land. (Lunar and Martian craters often show a central 'rebound' pinnacle, and it is possible that some explosion basins on Earth may have a secondary wave frozen into the floor of the basin. Such a configuration, if flooded to the right level, would appear as an island surrounded by water, further surrounded by a ring of land, then a second ring of water and, finally, the crater rim and general terrain.)

Cleito is reported to have delivered to Poseidon five sets of twin sons, and the land ruled by Poseidon ultimately was divided among them, except that the eldest son, Atlas (Atlas, Atlantis?) was, by right of primogeniture, designated to be the succeeding king and to hold the central island of the domain. (10 sons – 10 tribes of Israel?)

The island, according to the historians, was delightfully equipped by Nature with the fortunate combination of springs which furnished both hot and cold fresh waters. (Such springs are found today in islands such as the Azores.) Further, the land was noted for its prolific growth of all types of vegetation, which in turn fostered an indigenous population of grazing animals that included the elephant and horse. There were, of course, lakes, marshes, rivers, hills and plains, 'whatever fragrant things there are in the Earth . . . grew and thrived in that land.'

There must have also been great mineral wealth because the Atlanteans built for Poseidon a temple with pinnacles covered by silver and gold, a roof (ceiling?) of ivory, pillars and walls adorned with gold, silver and 'orichalacum'. Orichalacum, according to the historians, was a 'red' metal which could be dug from the Earth in many places throughout the land and, with the possible exception of silver and gold, was esteemed the most precious of metals among the men of that period. (If the red metal is copper, the reason for its high esteem becomes obvious. When mixed with tin and other elements, it becomes the bronze which was so widely used for tools and weapons before the more plentiful and therefore cheaper iron replaced it. Even today, especially among those who wish their artwork or craftsmanship to endure, bronze is considered a far superior and useful metal to any of the others, including gold and silver, because of its twin qualities of toughness and high resistance to corrosion.)

The Atlanteans are also reported to have dug a canal '100 feet wide and 100 feet deep' (probably 10 feet deep unless they had some other purpose than provision for the passage of watercraft) that extended from the 'sea to the hill'. The result of this effort made the kingdom famous for its magnificent harbor, which consisted of an inner and outer basin, with the usual assemblage of wharves, docks and marine supplies.

In another work by Plato, *Timaeus,* he reports that the Egyptians told Solon that 'Atlantis was no smaller than Africa and Asia joined together, an island set in the Atlantic not far from the Pillars of Hercules [Straits of Gibraltar], beyond which lay other fruit bearing islands, and beyond these an unknown continent which surrounds the true ocean [the Pacific Ocean?].' They said that the Atlanteans had partially conquered the unknown continent to the west, and had penetrated eastward into Europe, Italy and Egypt. The Athenians had defeated the invasion, however, and very 'soon afterward, Atlantic was engulfed by the sea'. This, according to the Egyptians, occurred 9000 years before the time of Solon. (One set of circumstances that would surely drive the Atlanteans to attack their neighbors both east and west would be the dreadful knowledge that their kingdom was in the process of being drowned.)

The Atlanteans probably felt the invulnerability of those who are long associated with danger. As the sea crept upward, they turned their city walls into dikes to keep out the water. They probably also invoked the aid of Poseidon. The faithful probably believed that Poseidon would not allow his sea to overrun his own fair city. As the waters rose higher and higher, they built the dikes higher and higher until the whole flat basin of the city lay below sea level. Meanwhile, they continued their normal lives, just as do the modern inhabitants of Los Angeles in the face of certain earthquake disaster forecasts.

According to Plato, a great earthquake and flood devastated Atlantis. An earthquake could have broken the walls that held the sea back. The water then poured through the breach, inundating and destroying the city and the Atlantean civilization. When the sea reclaimed the explosion basin that held Atlantis, it also drowned much of the physical evidence that arose there.

However, there are many indications that the culture was not totally destroyed since some sort of communication existed across what is now the Atlantic Ocean. For example, there are interesting architectural similarities from supposedly different origins. The walls built by both the American ancients and the Egyptians of the same era slanted inward at the tops, and their doorways were narrower at the top than at the bottom. Obelisks in Egypt are very similar to the columns found in Central America. Burial mounds were concurrently used in Europe, Asia and the Americas. On both sides of the Atlantic the manufacture and use of cement for concrete was employed. The Egyptians made sun-dried brick using straw as a binder while the Mound Builders of North America were doing the same. The pyramids of Egypt, Assyria and Phoenicia had their counterparts in Mexico and Central America. Furthermore, these Egyptian pyramids, the temples of the Etruscans and most of the pyramids in Mexico are constructed to line up with the cardinal points of the compass, north-south-east-west, which calls upon a similar knowledge of astronomy.

There is no evidence that Egyptian civilization developed within the Valley of the Nile. The record shows that 'instantaneously' the Egyptians were highly skilled and strong, and for centuries dominated their neighbors who were much less knowledgeable. One hypothesis that fits such an emergence is that they, perhaps as neighbors of the Atlanteans, first moved uphill and out of the Atlantic basin as the melting ice sheets uncovered fertile new lands.

The weapons on both sides of the Atlantic were similar at the same points in time. One interesting exception was the spear-throwing accessory used by the Americans. It was simply a wooden rod with a socket on the side at one end. The butt end of a spear was placed in this socket, while the other end of the rod was held in the hand. When the assembly was impelled forward by the thrower, the rod became an extension of length of the thrower's arm and thus the spear could be projected with greater velocity and to greater distances than the opponents' projectiles. The name of this assembly was the 'atlatl' (Atlas, atlatl, Atlantis, Atlantic).

Many religious beliefs were similar. The Aztecs, the Gaunches of the Canary Islands, the Egyptians and the Peruvians all believed in the immortality of the soul and the resurrection of the body; thus they all preserved their dead by embalming them. The Hebrews, Phoenicians, Egyptians, Asori, Hondurans and Central Americans all practiced circumcision. The Peruvians, Chaldeans and Egyptians divided the year into twelve months and the months into seven-day weeks, and added extra days to give the year 365 (and a quarter) days.

A very old, unnatural and unique custom practiced among the Iberians of northern Spain, the Basques and Corsicans of France, the Tibereni and the people of both North and South America is the act of 'couvade'. The literal meaning of the word is coward; the custom demands that when a child is born, the father 'takes to his bed as if he himself had suffered the pains of childbirth, cares for the child, and submits himself to fasting, purification, or various taboos'.

The popular theory about the origin of humanity on the continents of North and South America has all of these people being descended from migrants who crossed the Bering Sea land bridges when the oceans were depleted by ice sheets. It is strange, then, that while the descendants of the East Asian peoples have a 30 to 60 percent incidence of blood type B, many Amerindian tribes are composed of persons without, or with only a minute number having, this B type blood. Hematologists believe there is little likelihood that people with these very different percentages of similar blood types could be descended from the same ancestors.

The language of the Berbers of Africa is more closely related to that of the residents of the Canary Islands than to that of their surrounding neighbors on the continent. The name 'Canary' for these islands comes not from the bird, but from the Latin *canariae,* which denotes the great numbers of dogs that populated the islands at one time. The Basques have no lingual affinities with any other European race, but their language is similar to the language of the

Amerindians. The Basques have lived since before recorded history in the mountains and seacoasts of the Bay of Biscay: the junction between France and Spain that also is the upstream region of a submarine canyon which extends down to the floor of the Atlantic. The Basques also show a lower incidence of blood type B than any other people in Europe, another similarity to the Amerindians. Both of these distinctive groups of people, as well as the residents of the Canaries, may have been neighbors or the descendants of the Atlanteans, along with the Amerindians to the west. It is likely that in the event of catastrophic dislocation of a civilization, such as the one proposed for the Atlanteans by their being flooded out and fighting with 'continentals' for living space, domesticated animals like the dog would be left to fend for themselves. Perhaps those dogs that neither reached the safety of continents before the floods nor were drowned along the way are represented by the animals that later caused their sanctuary to be named the 'Dogs Islands' (*Canariae insulae*).

When Columbus arrived at the easternmost part of the American continental shelf, the island in the Bahamas that he named San Salvador, he was amazed to find a well-developed society of people there. Despite the fact that he had bent all of his efforts for years on proving that he could reach India by this westward route, his journal entry reflected a completely different idea. Instead of writing that the people he had encountered were, as he had long proposed, neither white nor black but remarkably like the Indians he sought, he made an entirely different observation. He wrote that these people were neither white nor black but of the bronzed color of the natives of the Canary Islands.

This undoubtedly true statement was very unlikely to bolster his goals in the Royal Court at Spain – so it stands out as a doubly significant observation.

Before the traffic in ships brought corruption to the characteristics of the people involved, we must assume that the native populations of these island chains, separated today by 3000-plus miles of ocean, were remarkably like each other, but not like the Africans or Europeans.

Dr Maurice Ewing of Columbia University recently announced that after 13 years of exploring the Mid-Atlantic Ridge, searching for lost cities, he had found no trace of any. Sounding, dredging and underwater photography have produced not the slightest results. But is this conclusive evidence that Atlantis does not exist?

The action of the surf rolling over the sunken city of Atlantis as the sea continued to rise would have literally leveled the city. There is no reason to suspect that any building would have been left standing to be located by scientists today. Worms would quickly destroy wood, straw and any other organic building materials. And soon the few hard remains would be covered with mud and silt, making them invisible to the camera or to the eye.

Even if the location of the city were known, probes and cameras might not disclose its ruins. A coring probe that fell upon any building material that was tough enough to survive on the ocean floor would only be deflected, and perhaps broken or bent. Landing on a bronze lintel, a building block, or a cobblestone would only produce an incomplete core – of which many are found.

Nor is undersea photography the answer. Under the best of circumstances and with brilliant lighting, a camera at the depth of three miles might have a range of perhaps 50 feet. Under less than perfect conditions, the range is less than 10. Locating the rubble of a city buried under yards of mud and silt in those conditions is like trying to locate a razed and buried Peoria, Illinois, by cruising over the Midwest on a cloudy and foggy night in a dirigible, trailing an Instamatic camera on a three-mile-long string.

Despite these difficulties, the search for Atlantis on the Mid-Atlantic Ridge may still have some hopes. The United States government has highly detailed maps of the floor of the Atlantic that could be of great help in locating possible sites for Atlantis. Although Plato's description of the city may be fanciful, an examination of these now governmentally 'classified' charts may reveal locations that would be congenial to the development of a great city. Armed with the information on these charts, the locations for the search for Atlantis may be narrowed down, and the city itself may be located.

The present-day migratory behavior of certain birds, eels and fishes has long puzzled naturalists who observe that they follow illogical routes. There are several species of birds that fly great distances over the submerged Mid-Atlantic Ridge of the Atlantic Ocean when safety, food and resting places are available along alternate continental routes that are not significantly longer. Salmon have been traced traveling between their places of birth, high above the present sea level in river tributaries, to regions of the oceans which overlie the very deepest basins, a location that does not provide the prolific food supply of the continental shelves where they normally feed. Similar is the behavior of the Atlantic eel which populates the fresh-water streams and lakes on both sides of the Atlantic until the time arrives for mating and spawning. Then, in response to a mysterious urge, these adult male and female eels move with a single accord to the Sargasso Sea, a portion of the western Atlantic that lies over its deep, large basin between Bermuda and Puerto Rico. There the young are born, and in due time they, just as mysteriously, obey another instinct which leads them back to the lake or stream from which their parents migrated earlier.

Viewing these mindless responses to a 'program' whose origin is not discernible now, it is reasonable to conclude that anything as powerful and unswerving as the direction and goals of these migrations must have become a fixed pattern for good reason at some earlier time. Looking at the potential hazards to migratory animals in a world that periodically trades its oceans for ice sheets and vice versa, the reason quickly asserts itself. Great sea level changes and the accompanying climatic and vegetation growth disruptions would gradually winnow out all routes that were not capable of supporting life for the animal involved along the entire length of the route. To take the simplest example: throughout the countless generations of eels and salmon, whether the oceans were at high or low sea levels, whether the land was covered by ice sheets or not, there was always a continuous stream of fresh water between the deepest points in the oceans and the several spawning or growing locations on the continents somewhere along the present migration routes.

Thus today's survivors are merely obeying an instruction that has proved in the past to cause their antecedents to survive, but has eliminated from the scene those families that were not fortunate enough to have followed the 'permanent' water routes and were either extinguished or became trapped in basins where their migratory habits were obliterated by time.

The Biblical story of the Great Deluge is well known. Less known is the fact that not only do the Hebrews, Greeks and Babylonians have this story in their histories, but also the Aztecs, other Amerindians and Southeast Asians and Chinese. It is commonly assumed that all of these people are simply responding to the last retreat of the Pleistocene ice sheets, in which uniformitarians grudgingly allow that worldwide sea levels may have risen as much as 600 feet to the present levels of today. The story of Noah's problems does not fit that analysis. Had Noah been faced with a slow, 600-foot rise in sea level at the rate it takes ice sheets to elevate the last 600 feet of the oceans of the world, he probably could simply have walked his troupe slowly uphill to higher ground.

Nor is it likely that he, or the historians of the other cultures around the world, would have judged this slow change of scenery to be of sufficient importance to document the incident in great detail. The day-to-day vicissitudes of life are taken, by all of us, to include problems such as fire, flood, famine and war with only passing note made of their occurrence. If the flood story was noteworthy because of a

600-foot elevation, the details should have been vastly different for each historian involved, simply because the terrain upon which each civilization was located would force different effects to be noted. Some societies, located on islands, would have been completely lost if the oceans inundated the entire land. Others, living upon slopes that continued upland to equally tenable living areas, would take little or no notice of such an occurrence.

Given, however, the circumstances that all societies were forced to live upon the floors of what are now the oceans because the high-altitude continents were covered by massive ice sheets, the changes forced by the return of the oceans to their beds, no matter how slowly such took place, would result in horizontal relocations of entire societies for thousands of miles, complete disruption of the ways of living and untold horror stories of those who stayed and prayed for the cessation of the waters and were never again seen – then it is reasonable that an identical story of a deluge that ended the known world became widespread.

Uniformitarians, insisting that only minor ocean level changes have occurred, place the location of Atlantis in the Mediterranean and attribute to the Minoans the source of the legend. They choose to ignore the dimensions of the land of the Atlanteans as being 'larger than Africa and Asia together', or perhaps consider such claims to be historical bombast. Therefore, the search for Atlantis has turned to the present-day continental shelves and shallow sea floors, on the assumption that the descriptions of Atlantis are grossly exaggerated and that the lost city is probably just a location near present shorelines.

As this is being written, the famous Jacques Cousteau and his crew of the ship *Calypso* are being outfitted in Marseilles, France, for an 18-month search for Atlantis. Their expedition, in keeping with the uniformitarian view of the 600-foot limit of sea level change, will search in the vicinity of the Greek island of Santorin, north of Crete. They will probably find the evidence which they seek: that humans have inhabited the continental shelves and Mediterranean

Sea floor before moving up to the present continental lands in response to rising sea levels.

Among the thousands of books and articles written about Atlantis may be found many citations of the evidence of human habitation on the continental shelves. Unfortunately, this evidence has been misinterpreted in a very important respect. It is not evidence of the lowest altitude in the ocean beds to which humanity was forced. Instead, it should be recognized as the proof that the people of that time were still retreating from the rising seas, which had, for centuries, driven them off the ocean floors, ever higher onto the continents.

University of Miami marine scientists made several significant announcements in 1975. One is that they have incontrovertible proof that fossils and limestones taken right off the surface of the sea floor 3000 feet below the present – and thousands of miles from any present-day shore, on the submerged Mid-Atlantic Ridge – contain substantial amounts of rainwater, indicating that the land at one time was above the sea surface. Another is that there was definitely a continent between the African and American land masses and that the Atlantic Ocean at that time was as small as the Red Sea. And most important is the evaluation that 11,500 years ago – which is right on the mark for Plato's dating – a sudden flood of icy waters caused the deep-sea creatures to change their characteristics so sharply that this change serves as a time-line in cores of sediments all over the world.

Hopefully, the Cousteau expedition's success in finding interesting human artifacts within the 600-foot region will not divert efforts to continue the search. In view of the many items that have been listed here arguing for a much larger, more important society and a location from which European, African and American cultures could be affected, the basin of the Atlantic Ocean seems to be a much more logical site.

In broad terms the topography of the floor of that basin, to the meager extent it has been charted to date, shows that a three-mile reduction in sea level would expose continuous land from Europe to North America and from South America to Africa, with similar exposure in the Indian and Pacific Oceans. Only the very deep basins such as those underlying the Sargasso Sea, and some smaller areas near other continental slopes, would contain oceans in the North Atlantic region.

Since the Mid-Atlantic Ridge, presently a submerged mountain range, would be a substantial watershed and topographical divide at lowered sea levels, it would not seem to fit the concentric-circle land suggested by Solon for Poseidon's Atlantis. However, on either side of the ridge may be found basins of sufficient area to accommodate substantial lands and sea, and both appear to be extremely large meteoroid-impact basins. Undoubtedly, detailed charting of these regions will reveal that, as seen on the Moon and Mars, large basins often embrace smaller craters that represent subsequent meteoroid impacts.

Taken on the premise that Egyptians, Greeks and other Afro-European societies who fought with them successfully were not the descendants of the Atlanteans, but that Berbers, Basques and the vast numbers of North and South Amerindians could very well be, it appears probable that the greater number of Atlanteans migrated west to the Americas. That, in turn, suggests that the western basin of the Atlantic was the site of Atlantis, rather than the eastern. Within that western basin, still in the broad terms dictated by the lack of detailed charts, there appear to be four, possibly five, major rivers which would flow into the hypothetical land of Atlantis. From the northeast, flowing in the well-known Mid-Ocean Canyon that has been traced from both sides of Greenland, would be a magnificent river that drains northeastern Canada, Greenland, Iceland and parts of the land between Iceland and Scotland. From the beds of the St Lawrence and Hudson Rivers would gush lesser, but still substantial volumes of water. From the west or southwest, certainly from the grand waterfall of the Bahama Island Shelf, would empty the waters coming from ice overlying the Gulf of Mexico and the Caribbean Sea. Finally, from the south or slightly southeast, there would be a river draining northeastern South America and the remainder of the Caribbean Sea.

Presumably it would be the confluence of one or more of these rivers with the low-sea-level ocean that would become the site of economic power, and therefore by derivation, the site of political power and the Lost City of Atlantis. A survey that seeks traces of a geometric pattern in any of these several regions might have startling success.

Water Ages and Ice Ages

Dr Cesare Emiliani, of the University of Miami's Rosensteil School of Marine and Atmospheric Science, analyzes seafloor cores containing ancient seashells to determine the temperature of the ocean at the time that those seashells were formed. His researches have led him to propose that in the last 400,000 years there have been eight eras of bitter cold, and about thirty less dramatic fluctuations. This is a radical departure from the classical view that Earth experienced only four ice ages in the last million years, each lasting 100,000 years and separated by intervals at least that long.

Dr Emiliani's research also disclosed that there were seven eras of 'raging heat' – in the equatorial regions – during the Earth's recent history, along with thirty smaller variations which presumably had periods of higher temperature than exist presently on Earth. These periods of high temperature caused an effect opposite to that of an ice age. In an ice age, considerable portions of Earth's water lies on the continents in the form of ice sheets. During the time of higher than normal temperatures, the polar ice caps and continental ice sheets melt away and raise sea level above its present range. Since the eras when this happens are the opposites of the ice ages, they may be aptly named 'water ages'.

Since the ice sheets of the last ice age are still retreating, it is possible that we are approaching the peak of a water age. Water ages in the past have had profound effects upon the land. The seas covered vast areas of all the continents, reducing them to low islands separated by shallow gulfs and bays. The Middle West of the United States was frequently submerged. Europe lost Holland and Germany, and many parts of Africa, Asia and Australia found their continental margins drowned. Vast beds of soft sea-bottom sediments were laid down on the continents at these times.

This rising of sea level, although caused by warm temperatures, would itself cause climatic changes. In effect, the rising of sea level during a water age lowers the altitudes of the continents. Mountains present a lower profile relative to sea level; shallow, solar-heat-absorbing seas are spread over the continents and worldwide temperatures are raised. The climate also becomes more uniform because the increased area under water permits a greater transfer of heat from the Equator to the poles by water flow. The lower mountain profile permits a greater air transfer of heat also, not only between the Equator and the poles, but also between the sea and land.

The combined effect of the warming of the Earth and the lessening of climatic differences between Equator and the poles can account for some of the unusual distribution of fossils which are found today. Under these circumstances, it is easier to understand the finding of hippopotamus bones in Great Britain, fossilized corals within the Arctic Circle, and remains of the giant sequoia found north of Labrador. In addition, the shrinking of land areas during water ages would have caused mass migrations and unusual herdings on high ground. Those areas still above water would temporarily hold animal populations much denser than we see today. Any deposits of fossils that occurred during these ages would thus be richer and more varied than fossils deposited under other climatic conditions.

A cavern found in 1912 near Cumberland, Maryland, contained fossils similar to animals which inhabit the area today. But it also contained fossils of animals that now live in entirely different climates, both to the north and to the south. Similarly, the Norfolk forest-bed in Britain lies in close association with fossils of arctic plants, freshwater and marine plants and shells and animals that are now associated with tropical climes. Surprising as it may seem, each of the animals and plants found in these deposits could have been a bona fide native of the area at the time that the climatic fluctuations of the world made that particular area hospitable for that particular life form.

During water ages, high temperatures have melted the polar ice caps and sent life forms scurrying to shrinking areas of high ground in order to survive. Conversely, during ice ages, the continents have all been frozen and the oceans have fallen for miles, forcing life to retreat from the highlands and live on areas that were formerly sea floor. When those ice ages ended, they again returned but perhaps to different continents.

Today, research is being done on the sea floors with tools and techniques unknown just a few years ago. The first evidence gathered through these new techniques indicates that the deep sea floors have limestone beds, much to the confusion of the adherents of the drifting continents theory. Future evidence, drawn from cores driven hundreds of feet into the sea floor, will disclose that the sea floors are very ancient structures. Layers of limestone, separated by abyssal ooze, will finally yield the continuous record of life so long sought by paleontologists. Further, these cores will also show that high or low sea levels have existed all over the world simultaneously, strongly supporting this proposal that worldwide ice ages have been the cause of these changes.

7

The Ice Ages

Evidence of Past Ice Sheets

The writings of geologists are often subdued and difficult to appreciate. Perhaps they tend to obscure their subjects because the very enormity of them tends to humble mankind's greatest achievements. Take, for example, the objects that are variously called decollements, wildflysch, thrust faults and erratics. The French term *décollement* means 'unglued' while flysch is a type of Alpine rock and wildflysch is such rock when it has moved miles from its point of origin in slabs and folds of mountainous proportions. Where an entire mountain range, hundreds of miles long, has been displaced horizontally by some unknown force, the plane upon which it moved may be conservatively noted as a thrust fault. And the mountain-sized masses that moved are sometimes known as decollements. But where the objects are of medium dimensions, and the force that moved them has been agreed upon the jargon is more descriptive, and such masses are known as 'glacial erratics', because it is accepted that they are moved by the ice sheets.

Thirty miles south of Calgary, Alberta, stand two huge quartzite boulders which are jointly known as the Ototoks Erratic. This mass, like all erratics, is totally unrelated to its surroundings. It does not resemble the rock on which it lies, nor is there any source for quartzite in the vicinity. These rocks have been moved at least 50 miles. Together they weigh over 18,000 tons.

The same force that moved these gigantic boulders also transported a huge mass of chalk to Malmo, in southern

Sweden. The massif is 3 miles long, 1000 feet wide, and from 100 to 200 feet in thickness. It has been transported a great distance and is the site of a limestone quarry.

A similar massif erratic on the eastern coast of England has had an entire village built upon it.

Uniformitarian purists concede that erratics are the result of the actions of continental glaciers which, during the ice ages, entrained massifs, boulders, pebbles, dust and materials of intermediate sizes, and carried them away. And they concede that when the ice melted, the materials were left where they had accumulated, often many hundreds of miles from their native sites.

But there are also erratics that lie in more equatorial regions. These erratics are far south of the line which uniformitarians recognize as the southern limit of the North American ice sheets. For example, there is a scattering of granite boulders in southern Georgia, extending down to Dothan, in the southeast corner of Alabama. Here, unlike their northern counterparts, uniformitarians call them 'outcrops', which evades the need to explain how these truck-sized boulders got so far south.

Furthermore, in Florida there are many erratics, either very small or hidden from view. The Florida peninsula, from Perry to Key West, is primarily a limestone sandwich that covers igneous rocks (rocks of volcanic origin) one to two miles below. This igneous material represents buried mountains which today do not extend near the surface anywhere.

The natural decomposition of the limestone cover yields a soil that has no igneous material, either in the form of rocks or as silica clays and sands. Yet Florida has long inland beds of sand which appear on the surface as prehistoric dunes and beaches. Other deposits of silica sand have been found when excavations and drilling have probed into the limestone below. Pockets of kaolin, a clay derived from igneous sources, have also been extensively mined in Florida.

Wildcatters seeking oil in Florida have drilled through

many layers of successively older limestones, down to rocks formed during the Cretaceous Period, when dinosaurs ruled the Earth. On the way down, their drill stems have frequently been snapped, much to the disgust of the drill crews. The damage has been attributed to encounters with deeply buried granite boulders, which are very hard relative to the limestone. Only a few hundred holes have been drilled in Florida, yet the incidence of damage to drilling stems prompted one angry driller to swear that there were more boulders in Florida rock than there are 'raisins in a loaf of high-priced raisin bread'.

All of these boulders must be classified as erratics, since they are unrelated to the limestone in which they are found. Furthermore, their occurrence at many depths means that the force that brought them there has acted many times in past geological eras.

Hematite, or iron ore, is a middle-sized form of erratic rock in Florida. A well-known tourist attraction in the center of the state is Bok Tower, built upon a 'mountain' – which in lowland Florida is only 325 feet high. The tower is named for a philanthropist who appreciated the church-like quiet of this peaceful knoll and made it available for all to enjoy. The tower and its surrounding garden of trees and plants rests on a mound of silica sand and decomposed iron ore that is not derived from the limestone underneath.

A few miles south of Bok Tower, northwest of Lake Okeechobee, there is a cut through a hill which gives the highway a more gradual descent. This cut was dug through another deposit of silica sand, again unrelated to the underlying limestone. Within this deposit, cobbles of iron ore still in rock form may be found. There is also much evidence of decomposed iron ore, along with ancient shells and chunks of coral.

The rough surfaces and sharp corners of the iron ore rocks conflict with the uniformitarian proposition that these deposits were, bit by bit and over vast periods of time, washed down to Florida from their area of origin in northern Georgia. This migration is supposed to have taken place as the peninsula of Florida slowly rose and sank beneath the waves. If it were true, surely the iron ore, instead of being found in the form of rocks, would have been ground into an unrecognizable powder during the 600-mile journey.

Also, if the silica sand that is found in Florida was carried there from the mountains of Georgia by the actions of surf and wind, one would expect that the modern Gulf Coast beaches would be covered with silica sand. Instead, the beaches, from Clearwater to the Keys, are made of ground-up fragments of the ancient limestone and modern sea shells. Yet directly inland there are beds of prehistoric silica sand being mined for use in building concretes.

Thus geologists are confronted with a spectrum of objects ranging from mountains on the one extreme down to mysterious deposits of silica sands and dusts on the other. The difficulty of identifying the forces that transported these objects is that at present only certain temperate zone regions of Earth's surface are accepted as being the sites of recent ice age activity, and the certain evidence of Equatorial Zone ice age activity in the more distant past has been discounted by those who embrace the continental drift hypothesis.

So they classify only the middle-sized objects that are clearly located in 'safe' ice-sheet regions as 'glacial erratics', while awaiting further ideas about Earth's past to explain the dusts and mountain movements from the poles to the Equator.

Ice Age Causes

That the Earth has seen ice ages in the past is a fact which is no longer debated. Ever since Louis Agassiz made his presentation before the Geological Society of London, on November 4, 1840, the body of evidence pointing to the repeated formation of continental ice sheets has grown steadily. Refinements to the knowledge of ice ages have taken place, such as new findings regarding the number of ice ages or the extent of the ice sheets, but the fact of their existence is unquestioned.

Despite the universal agreement that there have been ice ages and despite the great amount of research that has taken place on the subject, we have not been able to agree upon a cause for them. Many hypotheses have been offered and debated, but none has yet become widely accepted. In this area of inquiry, the doctrine of absolute uniformity is an insurmountable barrier to the development of some hypotheses, because it insists that no sudden and catastrophic changes to Earth's surface have ever occurred.

Many of the ideas dealing with the beginning of ice ages have their roots in the amount of energy the Earth's surface receives from the Sun. Indeed, the Sun has a formidable effect on climatic conditions on Earth. A 30 percent increase in the amount of solar energy which the Earth receives would destroy all life on Earth. More important to this subject, a 13 percent decrease would turn its oceans to ice.

But what could cause such a change in the amount of solar energy the Earth receives? Some propose that the Sun's output of heat varies, and that this causes ice ages and water ages. Sir George Simpson has proposed just such a mechanism, but with a twist: he says that an increase in solar radiation can trigger an ice age. According to this idea, an increase in the Earth's temperature causes increased cloudiness and precipitation. The increased cloudiness protects polar ice sheets from melting while increased snowfall encourages their growth. Eventually, however, the increasing heat destroys the ice sheets. As the Sun's radiated energy declines, the process is reversed and another ice age occurs. Thus, according to Simpson, two ice ages occur with every fluctuation of the Sun.

Some have proposed that a great increase in sunspot activity could reduce the Sun's output of heat enough to trigger an ice age on Earth. Sunspot activity does cause a drop in the Earth's temperature – worldwide temperatures fall one or two degrees during times of high sunspot activity – but the pattern of solar storms is very regular. Sunspots increase and decrease in a predictable 11-year cycle. Since there seems to be no theoretical device to explain why this cycle would stall or run wild, most investigators have rejected sunspots as a cause of ice ages.

Dr Ernst J. Öpik, the University of Maryland astronomer, suggests that periodic expansion and contraction of the Sun changes its heat output and thus triggers ice ages on the Earth. Shrinkage of the Sun would increase the rate of fusion in the Sun's core, raising the temperature and the amount of heat which the Earth receives. Expansion allows the Sun to cool, causing a fall in the energy emitted and, thus, in the temperature of the Earth.

All proposals based on the variability of the Sun suffer from the same problem: the long-term variability of the Sun has never been established. Although experiments carried on since 1918 have indicated that the energy output of the Sun has decreased about 3 per cent in that time, the evidence is not conclusive. Therefore proposals that changes in the Sun's radiation of energy caused the ice ages can neither be accepted nor rejected at this time.

The motion of the Earth itself can change the amount of energy which it receives from the Sun and the areas which receive that energy. The most easily observed of these motions is the rotation of the Earth, which causes day and night.

In addition, although the Earth's angle of inclination (difference between the axis of rotation and the plane of revolution around the Sun) is presently 23½ degrees, this has not always been the same. Like most of the other motions of the Earth, the angle of inclination is slightly variable, oscillating on an apparently regular schedule. At times, the Earth is closer to vertical in relation to its plane of orbit, and at other times it is tipped at a greater angle.

All together, there are three motions of the Earth that can affect how much heat the hemisphere gets from the Sun: (1) the distance from the Sun, (2) the length of summer, (3) the angle of inclination. Each of these factors changes on a regular, predictable schedule. Although none of these is powerful enough to cause an ice age on its own, there is one proposal, called the Milankovich hypothesis, which suggests that their combined effects might be sufficient to bring about an ice age. That is, if a period of long winters combined with the Earth's being tipped farther away from the Sun in winter, with winter occurring at the Earth's farthest distance from the Sun, those winters might be much more extreme than those we see today. Ice could pile up during those bitter winters that couldn't be melted away by the shorter summers. The glaciers of the Northern Hemisphere might then expand, creating an ice age.

The Milankovich hypothesis requires that ice ages appear on a regular, predictable schedule and can be verified or rejected according to whether the history of the Earth reveals that ice ages showed up on schedule. When first introduced, the hypothesis was rejected because the observed facts did not fit the theoretical schedule. However, in recent years refinements have taken place in both geology and the Milankovich hypothesis which apparently bring them closer together.

But there are still three major problems with the Milankovich hypothesis. First, if it is correct, the Earth is now overdue for an ice age, and there is presently no sign of the beginning of one. Second, an ice age under these circumstances would take at least 15,000 years to come into existence. Third, the theory permits only the formation of glaciers in one hemisphere at a time, while geological evidence points to ice ages occurring all over the world simultaneously. While tempting the Milankovich hypothesis has rather crucial drawbacks.

Another hypothesis of the cause of ice ages is based on the Earth's use of the energy received from the Sun – not on any change in the amount of that energy. Adherents of this hypothesis point to the 'greenhouse effect' of carbon dioxide in the atmosphere. Much of the energy of the Sun comes to the Earth in the form of short waves, such as visible light and ultraviolet radiation. These short waves pass easily through the carbon dioxide of the atmosphere. After they reach the surface of the Earth, however, many of these rays are transformed into infrared radiation (heat), a longer wave which cannot penetrate the carbon dioxide to escape into space.

An idea that is based on the Earth's use of the Sun's heat is that of C. E. P. Brooks, which concludes that the world is always on the edge of an ice age because of conditions in the Arctic Circle. According to this hypothesis, a small drop in the temperature of the arctic air could start a snowfall of gigantic proportions. If the salt water of the Arctic Sea were to freeze, the entire area could be covered with snow. The high reflectivity (albedo) of this snow would reflect the Sun's rays, preventing the area from warming and melting. The warm water which the Gulf Stream brings to the North Atlantic would ensure that great evaporation of water in the area would feed the formation of clouds and snow. Like the snow, the clouds, too, would reflect the Sun's heat. The ice sheet, with a high pressure area above it and clouds of water vapor all round would have great, continuous snow storms around its edges which would feed its expansion.

Both the carbon dioxide and Brooks hypotheses suffer from the fact that they cannot be demonstrated. Each is technically interesting, but it is difficult to believe that either has the power to create the titanic ice sheets which once covered the globe. But there are other hypotheses which deal with causes that have had a definite effect on the Earth in the past.

One condition that has caused changes in the weather of Earth in the historical past has been the presence of dust in the atmosphere. In 1784, Benjamin Franklin postulated that the recent severe winter had been caused by volcanic dust shielding the Earth from the Sun's rays. And the 1883 explosion of the island of Krakatoa left dust in the upper atmosphere that took two years to fall.

Another striking study of the effects of high-altitude dust took place in 1912. An observer from the Smithsonian Institution in Algeria made measurements of the quantity of heat reaching the Earth from the Sun. On June 19, streaks of dust were observed along the horizon. In a few days the whole sky was filled with dust, which persisted for months. The phenomenon was observed worldwide, and was found to be caused by an eruption of Mt Katmai on the Alaskan peninsula. This eruption caused a 20 percent decrease in the amount of direct solar heat reaching the Earth's surface during the summer of 1912.

Since astronomers have concluded that a 13 percent decrease in solar heat reaching the Earth would cause the Earth's oceans to freeze, it is frightening to think what might have happened had the dust lasted more than a few months.

It has also been suggested that, besides volcanic dust, meteoric or cosmic dust may have been a cause of ice ages. Unfortunately, we know little about such dusts, so it is difficult to come to any conclusions regarding this idea. Certainly, if such dusts were present in the quantities that have been observed in volcanic dusts, the effects would be much the same.

Much is known about volcanic dusts, however, and some of this knowledge supports the idea of the ice ages being caused by volcanic action. Many of the particles which are ejected from an erupting volcano carry an electrical charge. This fact is demonstrated by the electrostatic discharges such as the spectacular lightning displays that often accompany an eruption.

A satellite picture taken of the volcano Tiatia in the Russian island of Kunashir shows a slower discharge of the electrical potential built up by the volcano's ejection of charged particles. This photograph, taken during the volcano's eruption in 1973, shows a solid cover of clouds moving toward the volcano. But the clouds are somehow prevented from reaching it, although they flow by on either side. It looks as if the clouds were cut away by a 32-mile-diameter 'cookie cutter' around the volcano. Some have suggested that the clear area is due to shock waves from explosions associated with the eruption, but the picture clearly shows that the smoke of the volcano drifts away, undistorted by any such percussions. The only other reasonable hypothesis is that the entire area has been electrified by the particles rushing out of the volcanic vent – which is a well-known effect on a laboratory scale.

Since not all of the charged particles ejected by a volcano lose their charge in lightning displays and such, dusts may be blown high into the stratosphere still retaining their charges. It is even possible that their charges may there be reinforced by the actions of wind and solar radiation. These charged particles will eventually return to settle on the surface of the Earth, but it could take hundreds of years.

The volcanic dust theory also has another attraction. It helps to explain the great snowfalls which must accompany the formation of continental ice sheets. Some have argued that cold is not enough to cause an ice age. If Earth merely became cold, they say, the oceans would simply freeze. There would be no evaporation to feed snowstorms, and continental ice sheets could never be formed.

However, if the dust in the atmosphere that shielded the Earth from the Sun's warming rays were accompanied by volcanic lava floods on the sea floor, the problem of the frozen oceans does not arise. The volcanoes would not only prevent the freezing of the oceans, the heat of those sea-floor eruptions might actually increase the temperature and thus the rate of evaporation of the oceans. With this combination of heating of the oceans by vulcanism and shading the Sun's rays by volcanic dusts, it is easy to see how an ice age might begin.

Like some of the other proposed causes of ice ages, the idea that they are caused by volcanic activity appears to suffer from insufficient power. There is vulcanism today, but it doesn't seem to be starting any ice ages. Vulcanism today, however, is substantially different from that which occurred in the past. Most vulcanism today takes the form of minor eruptions that form craters. There is only one example of the modern formation of a caldera, Krakatoa, which left dust in the atmosphere for two years. The Toba Depression in Sumatra, a caldera 100 times greater than Krakatoa's, might have sent aloft dusts that darkened the sky for centuries. Even small explosions like Krakatoa's could cause an ice age if they occurred frequently for a few centuries. The thousands of calderas that dot the Earth suggest that sustained explosions may well have occurred in the past.

All of the possible causes of ice ages that have been discussed so far have been based on a change in the amount of solar energy received by the Earth. Whether because of distance from the Sun and inclination of Earth, variance of the Sun or shading of the Earth, all have been predicated on the idea that the surface of the Earth received less heat from the Sun that it does today. But there is another body of proposals that does not depend on any change in solar energy at all.

One of these ideas is that the Earth has traveled through areas of space that had different temperatures, and that these temperatures affected the temperature of the Earth.

A number of propositions hold that there was no substantial change of the world's climate, but that the Earth's axis of rotation changed. This would cause the poles to be relocated, and thus the frozen polar areas also. Areas that were once temperate became frozen arctic wastelands, and areas that were polar tundra warmed and became congenial to life.

The presence of fossils of animals and plants in what are now lifeless polar areas makes this idea appealing, but, as textbooks point out, the energy required to change the Earth's axis of rotation is enormous. The elevation of ice or land has been proposed to have thrown the Earth way off-balance, thus greatly changing the location of the poles, but again, the elevation of all the mountain ranges plus Earth's water mass are insufficient to change the angle more than a few degrees.

It has been proposed, too, that the drifting of the continents might have produced the same effect. The problem is energy. As one investigator pointed out, geologists do not seem to realize the vastness of the Earth's size, or the enormous quantity of its momentum. When a mass of matter is in rotation about an axis, it cannot be made to rotate about a new one except by external force. Internal changes cannot alter the axis, only the distribution of the matter and motion around it. For the Earth to revolve about a new axis, every particle would have to change its direction.

The problem of a change of axis appears to be insurmountable. But even if this problem did not exist, a change of the poles would not account for the ice ages. A shift of the poles would only change the location of the polar ice. One area would freeze while another, formerly frozen, would thaw. But increasingly, evidence points to ice sheets covering all of the continental areas at the same time during ice ages. A shift of the poles would not explain a three-mile drop in sea level or the worldwide formation of canyons below present sea level caused by water returning to the oceans. It would not explain the fact that evidence of continental ice sheets is found all over the world, from poles to Equator.

Another proposition, called the ocean control hypothesis, suggests that ocean levels both begin and end ice ages in the Northern Hemisphere. This theory, proposed by Maurice Ewing and William Donn of Columbia University, suggests that high ocean levels permit warm water from the Atlantic to circulate into the region of the Arctic Ocean. This warm water evaporates and condenses in the northern latitudes, falling as snow within the Arctic Circle. This warm water flow continues to feed the glaciers until falling sea levels cut off the flow of the warm Atlantic water into the Arctic Ocean. With the halting of that flow, the Arctic Ocean freezes, evaporation halts, and so does the growth of the ice sheets. Normal solar heat then melts the ice sheets, causing sea levels to rise and permitting the warm Atlantic waters to once again flow into the Arctic Ocean, beginning the cycle again.

A variation on this hypothesis was put forth by William Lee Stokes of the University of Utah. In his view, the ice ages began when the isthmus of Panama rose to divide the Atlantic Ocean from the Pacific Ocean. Prior to that time, he says, the Gulf Stream flowed westward directly into the Pacific Ocean. With the rise of that isthmus, the Gulf Stream was redirected into the North Atlantic, causing warming and evaporation of the ocean in that area. This triggered the cyclical ice ages which Ewing and Donn described.

These ideas of ocean control have a self-limiting factor. They are designed to explain ice sheets that cover only high northern latitudes. But the evidence of ice sheets that is found all over the world indicates that, because of that self-limiting factor, these ideas must be rejected or substantially modified.

Another hypothesis of the cause of ice ages is increasing continentality, or the uplift of mountains. This idea, first proposed by C. E. P. Brooks, is based on the fact that periods of mountain building are closely associated with ice ages. Brooks concluded that when continents and mountains are raised, the flow of air and water from the Equator to the poles is decreased. Continents re-route and interrupt ocean currents and mountains interfere with airflow and jet streams, so that polar and equatorial regions are more isolated from each other. Also, the upper altitudes of the newly formed mountains may lie above the snow line, encouraging the formation of glaciers. With the reduction of the flow of warm equatorial air and water to the polar areas, the polar temperatures fall and ice sheets begin to form. These continue to grow until erosion lowers the new highlands and permits warm tropical air and water to flow to the poles and begin to melt them. This hypothesis relies upon either isostasy or plate tectonics to provide the 'increasing continentality'. As such, it describes an effect (which is disproved by this book) rather than proposing a cause.

Not only do theoreticians disagree on the forces which initiate an ice age, they also disagree on the speed with which it arrives. Most ice age hypotheses envision the proximate cause of an ice age to be the uncontrolled expansion of glaciers. Instead of their flow increasing during the winter and decreasing or stopping completely in the summer, the glaciers continue to flow all year long, eventually joining to cover vast areas. Slowly, the glaciers then begin to flow south, grinding and polishing as they go.

On the other hand, some propose that the continental ice sheets of southern regions did not necessarily flow down from the north. Instead, they may have formed in place in what has been termed a 'snow blitz'. Hubert H. Lamb and John H. Woodroffe have explained that in a snow blitz, ice sheets grow from the bottom up, on the plains as well as in the mountains. The ice sheets come directly from the skies, in the form of winter snows that fail to melt in the following summers. Even a few inches of unmelted snow persisting from one winter to the next is the beginning of an ice sheet.

The idea of a snow blitz is given further credence by the work of Nicholas Shackleton of the University of Cambridge. The atomic structure of the shells of small marine fossils gives a clue to the temperature of the world at the time they were formed. Based on his research, Shackleton concludes that at the point in the Earth's history he analyzed, ice sheets must have been growing thicker at a rate of 18 inches a year, across huge areas of North America and Europe.

Even this specific proof of the rate of ice accumulation is conservative, however, when the present rate of evaporation of the oceans is considered. That rate would permit the accumulation of a one-mile-thick ice sheet on all the continents in about 175 years. If undersea volcanic activity accelerated that rate of evaporation, the ice could accumulate even faster.

An Unusual Ice Age Mechanism

Most of the hypotheses of the causes of ice ages advanced in this chapter have drawbacks, and some have very serious deficiencies. Nearly all suffer from the same problem: the magnitude of the effect is greater than the magnitude of the cause. But there is one effect of an ice age that can cause it to encourage its own growth. C. E. P. Brooks almost hit upon it. Brooks noted the relationship between mountain building and ice ages, and made the supposition that high-altitude land increases could bring on ice sheets.

This is because, to an observer standing on the Earth, the oceans seem exceedingly deep and the atmosphere seems to extend a great distance into the sky. Compared to the size of the human body, the depths of the oceans and the heights of mountains are very great. But even though these altitude differences seem substantial to us, in terms of the size of the Earth they are insignificant.

If a ball bearing of one-inch diameter is machined to an inspection tolerance of one-thousandth of an inch, it will be rejected if it is found to be more than a thousandth of its diameter too large or too small. Assuming that sea level is the 'proper' size for the Earth, the greatest variance above that altitude, Mt Everest, the highest point on Earth, is approximately 28,500 feet above sea level. The deepest of ocean trenches are about 33,000 feet below sea level. Since the diameter of the Earth is approximately 42 million feet, compared to that diameter, Mt Everest is about 0·07 per cent above sea level, and the ocean trenches are about 0·08 percent below. Since the tolerance limits for the ball bearing were 0·10 percent, the Earth is smooth enough to pass this test for a ball bearing.

So the altitude differences that seem so great to us as human beings are really insignificant in global terms. The astronaut who described the Earth as looking like a 'big blue marble' was seeing the Earth in its true form. Instead of being misled by the exaggerations of mountains and deeps seen close to, he could see the true marblelike smoothness of the planet. But even though the difference between the highest mountain and the deepest ocean trench is slight, the range of altitudes that can support life is smaller still.

The highest of mountains can only support the sparsest of life because of the thin, cold air. Likewise, deep oceans are almost sterile because of pressures and the lack of oxygen at extreme depths. Nearly all of the world's life exists within a range of two miles of sea level. Above that in the air or below it in the ocean, the conditions are too rigorous to permit abundant life.

The cold of the upper atmosphere has a profound effect

upon the formation of continental ice sheets during an ice age. Due to the weight and compressibility of the Earth's atmosphere, 50 percent of the atmosphere is compressed to within 3·6 miles of sea level. A ride of 10 floors in a fast elevator is enough for one's ears to detect a change in the pressure of the air. The higher one goes in the atmosphere, the lower the pressure, and temperature falls at the same time.

On a hot summer day, an airport weather station may report temperatures of 90 degrees at ground level. But 6 miles above that weather station, the temperature of the air may be 60 degrees below zero. This 150-degree drop in temperature is due to the lowered pressure and density of the air at the higher altitude.

It is this fall in temperature that causes some mountains near the Equator to remain snow-capped all year round, regardless of the temperatures at lower altitudes. The altitude at which the temperature never rises above freezing is the 'snow line', and if a mountain exceeds this altitude, it will have a permanent cap of snow. This altitude varies according the climate of the Earth below it. In arctic and antarctic areas, the snow line is, in effect below sea level, because those areas stay frozen all the time. In temperate climates, such as the United States and Europe, the snow line is at an altitude of about 2 miles above sea level. Even in hot equatorial regions, the altitude of the snow line is less that 3 miles above sea level.

At present, only 2 to 3 percent of the Earth's area stands more than 2 miles above sea level. But what happens to sea level during an ice age? As the water from the oceans evaporates and is deposited on the continents in the form of snow, sea level falls. Every drop in sea level causes a corresponding descent of the snow line. More and more of the surface of the continents comes to lie above the snow line, not because the continents are rising, but because sea level, and with it the snow line, are falling.

When the sea level has fallen 1 mile, approximately *10 percent* of the Earth's surface is above the snow line. This rapid increase in high-altitude area has the effect of accelerating the onset of the ice age.

The fall of another mile in sea level means that suddenly 30 percent of the Earth's surface lies above the snow line. When sea level falls the full three miles, 40 percent of the Earth's surface lies above the snow line.

In equatorial regions, where the snow line had been at three miles above sea level, the snow line is lowered to what was formerly sea level. All of the areas that had formerly been land became frozen highlands. The thin layer of the climate zone that can support life moves down with the shrinking sea level to what had formerly been the floors of the oceans.

The effect of altitude on the formation of continental ice sheets makes the start of an ice age a dynamic process. Each step in that process reinforces the others. First, whatever triggers the ice age causes the beginning of an accumulation of snow in the continents. This accumulation of snow on land means that some water evaporating from the oceans is not returning to them, and thus sea level falls. As sea level falls, so does the snow line. As the snow line falls, an increasing area on the continents can catch and permanently hold water in the form of ice and snow. This expanding area of permanent snowfields reduces the amount of water which is returned to the ocean after being evaporated from it.

No matter what the cause of ice ages is, altitude effects accentuate it. When ice ages begin, they apparently rapidly shrink the oceans, sometimes all the way to the three-miles-below-present levels. Similarly, it appears that, much later, ice ages end suddenly, with great torrents of water roaring down steep canyons to refill the ocean basins. Both beginning and end are catastrophic, dwarfing the puny power of humans to change this world.

Past Ice Sheet Extent

Since the advent of the doctrine of absolute uniformity, geologists have been taught to think only in terms of millions of years about Earth's major changes. The idea of slow change and the rejection of any catastrophic explanation of a geological phenomenon are presently the cornerstones of the discipline.

The idea of slow change has extended into the philosophy as well as the reasoning of uniformitarians. When Lyell advanced the doctrine it was an attractive and reasonable alternative to theological catastrophism. Since then, evidence has been amassed which mandates its modification.

A sector which requires re-examination is that of the size of ice sheets of the past. The hypothesis that much of Europe was at one time covered with huge ice sheets became respectable about 1840. The conclusions which were drawn about the nature of those ice sheets were based on the research of Louis Agassiz, then a young Swiss naturalist. He performed his basic research on an alpine glacier, where he lived and noted its activities. Unfortunately, he did not emphasize that the action of a continental ice sheet could be vastly different from that of a glacier.

A glacier is a moving river of ice, fed by snowfalls in mountainous areas. It flows downhill and exhibits properties similar to a river, although its progress may be measured in feet per day rather than miles per hour.

To understand the properties of an ice sheet, we need only go to Greenland and Antarctica, where they still exist. On Greenland, ice exceeds 10,000 feet in thickness near the center of the ice sheet. Ice in Antarctica averages 7500 feet in thickness, and in Marie Byrd Land it stands over 2 miles.

It is important to note that these two ice sheets stand nearly motionless, except where they rest on downward sloping surfaces. Only there does the ice move perceptibly toward the oceans. If all the ice of Greenland were melted. the land left behind would be seen to be a low interior region surrounded by the ring of coastal mountains that now enclose the ice sheet.

If alien scientists were to land and look for signs of glaciation in a defrosted Greenland, they would find them only in openings of the mountainous coastal ring. These are the only places that would exhibit the striations associated with the scraping of glacier-borne rocks and the deposition of rock in lateral and terminal moraines. The interior, having had a relatively motionless ice sheet rest upon it, would show different signs of ice existence.

A good indication of the type of evidence which would be found after a standing ice sheet melted away was recently disclosed by the Institute of Polar Studies of Ohio State University. They took cores from Greenland's ice sheet – one was 4500 feet long – and from deep in the ice of western Antarctica. These cores contained snow and particles of material that had fallen on the surface of the Earth 14,000 years ago. A heavy concentration of particulate matter was found in this ice age snow, identified as the fallout from great volcanic activity plus sea-floor dust – at least 100 times heavier than present atmospheric dust levels. So it is apparent that if vast quantities of dust were tied up in ice sheets, and if those dusts were gently deposited by the ice as it slowly melted away, the evidence of the presence of an ice sheet would be a thick blanket of rich soil spread evenly over thousands of square miles.

That brings new significance to the soils covering the Earth's continental basins. They have that same mixture of volcanic and sea-floor dusts, spread evenly over hill and dale from the Poles to the Equator.

The absence of signs of glaciation in areas which have been covered by stationary ice sheets has led uniformitarians to some amazing conclusions about the locations of those ice sheets. In the United States, Pleistocene glaciation has been clearly *proved* to reach as far south as the 35th Parallel, a latitude which runs through North Carolina and close to Little Rock, Arkansas, and Los Angeles, California. Yet vast basins north of this line are claimed by textbooks to have been ice-free. These 'ice-free' areas supposedly include parts of Alaska and the far northern islands of Canada.

In the eastern hemisphere, the 35th Parallel runs near the Straits of Gibraltar, through the Mediterranean Sea to Israel, through the Himalayas, to southern Japan. Yet much of Europe and nearly all of Asia are proposed to have been free of ice in the last ice age. Even northern Siberia, today still a frozen land, is suggested to have been ice-free during the Pleistocene in most of its area.

Perhaps permafrost, an arctic phenomenon, can serve as a convincing indicator of the extent of the northern ice sheets. One form of permafrost, called 'muck' or 'frozen silt' in Alaska, includes vast deposits of trees and animal bones. Apparently this is a mixture of ground-together soil and ice. The other type is clear, clean ice which lies underneath the soils of the Siberian tundra. The two forms of permafrost form fixed layers which in some areas are as much as 5000 feet thick. The permafrost line in Siberia, north of which this layer never thaws, even in summer, reaches down to the 45th Parallel south of central Siberia, including the Gobi desert and its surrounding mountain ranges of Altai, Hangay, Great Khingan and Sikhote Alin, and then curves northward through western Siberia. In North America, the permafrost line cuts across Alaska at about the 62nd Parallel, then gradually drops southward as it crosses the interior of the continent to pass south of greater Hudson Bay, at about the 55th Parallel. It then ends at the Atlantic Ocean off Labrador.

Uniformitarian opinion on the origin of permafrost is that it is caused by a very low average annual temperature which prevents thawing of the ground, even in summer. This idea is refuted, however, by an examination of conditions of just two Siberian cities. The average annual temperature records for the cities of Novosibirsk and Khabarovsk are virtually identical, yet only Khabarovsk lies in the permafrost region. Clearly, a low average annual temperature is not enough to produce permafrost. Also, a low average annual temperature could only produce frozen soil, or 'muck'. The other type of permafrost, the clear ice, could never be produced merely by low temperatures.

Let us examine another hypothesis of the formation of the permafrost. Some parts of Siberia and Alaska were hosts to the slowly moving edges of giant continental ice sheets. These margins were glaciers that bore down on the ground beneath them with forces of more than a million pounds per square foot. The weight, combined with the slow flow of the ice across the land, ground and mixed soil and ice from the underside of the ice sheets together. Trees, rocks and animal remains all combined with the soil-ice mixture to produce the rich fossil deposits that are found in Siberia and Alaska today.

The clear permafrost, then, would appear to be the remains of the standing interior ice. It has since been covered with a layer of soil, probably by wind action, but the clear ice still remains beneath. Insulated by the soil and preserved by the continuing cold of the region, it has lain there for many thousands of years.

How does this hypothesis of the formation of permafrost compare with the facts? The permafrost line in North America is roughly parallel to the line which geology textbooks indicate is the southern limit of the Pleistocene ice sheets. That ice sheet line runs approximately along the 45th Parallel, to the south of the permafrost line, with the distance between the two lines roughly constant. This logically implies that the permafrost may lie in the area that formerly lay under very thick portions of the North America ice sheet.

In Siberia, however, the ice sheet and permafrost lines are wildly inconsistent. The textbook proposes that the Asian ice sheet reached down only to the 60th Parallel along most of its length in western and central Siberia, then curved north to the Arctic Ocean without including the coldest part of Siberia, its eastern third. This strange idea about the distribution of Pleistocene ice in Siberia undoubtedly is derived from a refusal to see permafrost as the remains of an ice sheet, which inevitably leads to the conclusion that the ice sheets of the last ice age were much more extensive than has been recognized.

It is difficult to imagine that ice covered nearly all of Canada without also covering all of its northernmost parts. It is similarly difficult to imagine that Siberia and Alaska were ice-free when large areas of the world south of them were covered with huge continental ice sheets. The absence of evidence of glaciation in these areas demonstrates that ice can exist in an area without moving. Ice sheets can stand still, miles thick over an area, and leave no mark of motion behind. When the ice age ends, the ice sheet can melt in place, and disappear as water. Since the ice sheet never moves, it leaves no sign of glaciation. Thus, large areas that seem to have no evidence of ice sheets should be examined for other signs of their presence. In 1865, Louis Agassiz found evidence of glaciation in equatorial Brazil. Similar evidence was found in the South American nation of British Guiana, and in equatorial Africa and Madagascar. These glaciers are attributed to the Permian ice age, much earlier than the latest one, but still illustrative of the areas that ice sheets cover. In the Northern Hemisphere, peninsular India, within 20 degrees of the Equator, was found to be the scene of past glaciation. There the ice flowed north, away from the Equator – but downhill on the continent.

The Antarctic ice cap covered practically all of southern Africa up to at least latitude 22 degrees and also spread to Madagascar during the Permian ice age.

Another ice age, during Paleozoic times, left similar remains. Continental ice sheets covered sections of South America, Africa, the Falkland Islands, India and Australia. In southwestern Africa, deposits related to these ancient glaciers are as much as 1800 feet thick. In many places, the now-lithified deposits (tillites) rest on older rocks striated and polished by these vanished glaciers. Evidence of widespread ancient ice sheets also comes to us from the Sahara desert. In 1961, a small group of French and Algerian geologists studied striations in Ordovician rock outcrops in eastern Algeria. Incredibly, the striations looked glacial! If they were glacial, it would have to be evidence of continental ice sheets.

For several years, the Institut Français du Petrole, and later the Institut Algérien du Petrole, sent in field party after field party. In neighboring Mauretania similar traces were discovered, and others have now been found all across North Africa. After extensive travels in the field, as well as air reconnaissance, the group was unanimous: there had been an ice sheet over the Sahara.

The presence of evidence of continental ice sheets in the Sahara has been explained by adherents of the drifting continents theory by saying that northern Africa was once in a much colder part of the world. One went so far as to say that the Sahara was once at the South Pole.

The finding of evidence of ancient ice ages in areas where such evidence is unexpected shows the extent that ice sheets can occupy. Some find the presence of ice sheets at the Equator inexplicable, but this is only because of the doctrine of uniformity. So great is their desire to retain the doctrine, they will move Africa to the South Pole rather than acknowledge that there could be a catastrophic change of the climate throughout the world.

These surprising evidences of ice sheets are not limited to remote parts of the world, nor are they relegated to the dim past of Earth's history. Signs of the last ice age have been found in North Carolina. The discoverers of these traces referred to them as 'the first positive evidence of glaciation south of New York' on the eastern coast. They discovered 30 to 40 half-inch parallel grooves averaging 12 inches long on an outcropping of rock on the northeast slope of Grandfather Mountain near Linville, North Carolina.

This proved that the glaciers existed in the North Carolina mountains 15,000 years ago and explained a mystery that geologists have considered for years – the finding of fossilized woolly mammoths, walruses and caribou, all arctic animals, in North Carolina.

It is significant that geology texts showed the southern limit of ice sheets to be about 500 miles north of Linville.

The cause of ice ages is unclear. Great volcanic activity providing both dusts to shield the Earth from the Sun's rays and heat to evaporate the oceans, seems to be a good possibility. This would also help to explain the deposits of volcanic ash associated with many Pleistocene fossils and also with other fossil remains. Whatever the initial cause, though, altitude effects speeded the formation of the ice.

The Polar Ice Caps

When the first Mariner probe and subsequent Viking orbiter close-up pictures of Mars arrived on Earth, they held investigators spellbound with the wealth of new information that they contained. Here, for the first time could be seen the topography of another planet, orbiting in the same solar system with Earth. Here we could begin unraveling some of the mysteries of what had happened to Earth by comparing its surface with Mars. One of the startling revelations concerned the polar ice caps.

Because of the disproportionate groupings of land masses on the planet, Earth's most visible polar ice cap is around the South Pole. The Northern Hemisphere – long called the land hemisphere by geographers and others – so surrounds its North Polar Ice Cap as to render indistinct where the ice cap leaves off and the continental ice sheets begin. For example, in global terms, the Greenland Ice Sheet is a vestigial remnant of the North Polar Ice Cap of the Pleistocene ice age.

But in the south, except for the minor effect of the land underlying Antarctica, we have an ice cap that is separate and distinct from the continental masses and their ice-age ice sheets. It is here that we may best understand the intensity of some of the past ice ages.

The South Polar Ice Cap of Earth must have grown both in thickness and in area during the ice ages. Its height is not easily ascertained, but its area may be easily estimated. During the Paleozoic ice ages, glaciers flowing out of the South Polar Ice Cap moved across parts of South America and Australia. This fact is known because the moving ice left striations and tills clearly showing its direction – moving up out of the ocean beds and onto the land.

By a conservative estimate then, the South Polar Ice Cap,

at that time, reached toward the Equator to a south latitude of about 35 degrees. Assuming that the North Polar Ice Cap had approximately the same dimensions – and we are encouraged in that belief by the positive evidence of Pleistocene ice sheets reaching down to the 35th Parallel in the United States – a photo of Earth from space would then show that only the middle third of the globe was not solidly covered with ice. In fact, given that the equatorial third has considerable evidence that its continental areas were ice-covered, the two polar ice caps would appear to merge over North and South America, over Africa and Eurasia and over Southeast Asia and Australia via Indonesia.

This of course must be the result when a 'water planet' such as ours ceases to enjoy the present-day warmth of the Sun.

Mars, on the other hand, is not a water planet. Its water supply seems to be only enough to provide small polar ice caps which are substantially diminished when the polar region tilts toward the Sun. The startling thing about the Martian ice caps was that they had a distinctive topography surrounding them. On Earth we have long, compressed ridges of surface material that we call mountain ranges. Later, you will see that Earth's mountain ranges are the result of the weight and expansion of ice sheets extruding soft continental materials.

On Mars there are also these compression-type ranges. They are found only surrounding the polar ice caps. Everywhere else on Mars may be found varieties of elevated land, but it is always clearly related to either volcanic action or the impact of planetoids and meteorites.

This is the kind of compelling evidence that is so eagerly sought for a new synthesis. Not only does the surface of Mars bolster the proposal for the universality of explosion basins on planets, but on the only other planet that has visible ice, the mountain ranges are formed only where the ice can accumulate in depth.

At present, approximately one-fourth of the Earth's surface is above sea level. Add to this the polar ice caps and lower the sea level slightly, and approximately one-third of the world quickly becomes land capable of holding continental ice sheets during an ice age. As sea level rapidly falls during the onset of an ice age, this one-third of the globe becomes chilled. Ice and snow are piled upon this land area.

The oceans continue to evaporate and pile their water on the continents as snow until they shrink to small seas. Where one-third of the Earth was land and two-thirds water, now those proportions are reversed. The one-third of the Earth that is now continental area becomes frozen highlands. The other third, the former sea bottom, is now the only area that can support life.

The shrinkage of the oceans to one-third of the Earth's surface, half of their former size, is caused by a three-mile fall in sea level. Evidence from the sea floor indicates that this low sea level persists for a substantial length of time, long enough to create a low-sea-level platform similar to the one which surrounds the present oceans, and long enough for the shallow seas to lay down a new floor of limestone.

But on the frozen continents, those highlands on which we now live, the water that once covered two-thirds of the Earth is now piled up on one-third of the Earth in the form of ice. The three-mile fall in sea level causes ice an average of *six miles thick* to cover all of the area that we now know as land.

8

Gifts of the Ice Ages

Petroleum

Around us today, life flourishes. Forests and grasslands, people and animals, cover the land. The lakes and oceans teem with life. Ever since life first appeared on Earth it has been this way. The forms of the plants and animals have changed, but the profusion of life – the total volume of organics – grows to the limit of the food supply. Life depends on food and sunlight. In the seas the food is suspended in the water. On land, the source of food is the soil. Where does this rich and nutritious material come from? And, besides petrification, what else happens to the plants and animals when an ice age clamps its grip on the continents?

An ice sheet forming over a body of water such as the Gulf of Mexico creates a capsule in which all the life forms are trapped and frozen. When organic matter and water are trapped together between impermeable layers, the decomposition that follows is performed by types of bacteria which produce oil and gas as their by-products. That this is true can be easily shown by experiment in which manure or garbage is held in a closed container to produce methane.

Until recently, most oil-genesis proposals carefully avoided any hint of past catastrophes, in keeping with uniformitarian teaching. Proposed instead were the possible ways that sea and lake bottom muck could, 'over periods of millions of years', somehow slowly enrich itself with vast quantities of microscopic organics, filter these organics down and out of its mud – leaving the latter as a clean shale – and gather the resultant crude oil into pockets from which lucky wildcatters might some day extract it. Only recently have the oil companies, in their TV ads, begun to imply that the past may not have been so calm and peaceful. They now urge conservation because, 'there aren't any more dinosaurs to replenish the supply'.

When Drake drilled for oil in Pennsylvania, he was not concerned with considerations of oil genesis. With straightforward commonsense, he reasoned that oil did exist where he sought it because it frequently appeared in the neighborhood farmer's water wells. It was no surprise to him, therefore, when he drilled deeper than farmers usually did, to find the expected 'petroleum' or 'rock oil'.

It was later, when others sought to improve upon his efforts by tapping the 'source' of the oil flow, that questions about its origin began to be discussed. The first consensus was that oil, like water, simply forms and flows out of rocks at certain places – just as springs seemed to do – and that consequently, oil also would make its way downhill but at some depth under the water streambeds. This idea, bizarre as it is today, was then acceptable because men in high places were arguing about whether or not the Earth was cave-ridden to its core, if not downright hollow inside. So they drilled near the creeks flowing away from Drake's well and were initially encouraged in their beliefs by frequently finding oil in these places. But their rationale was ultimately abandoned as more and more large strikes were made by unbelievers on the hilltops, while the creek and river bed deposits proved to be meager.

So another straightforward technique evolved – one that is employed successfully to this day – which disregards the source of oil and gas but coins the phrase 'oil is where you find it'. The proponents of this concept developed 'creekology' to locate their drilling sites. They read maps or tramped the countryside to find places where the local creeks and brooks all flowed away in a radiating pattern. Such places were rises on the land surface that sometimes covered underground rock domes. Such domes often contained vast quantities of oil and gas.

To modern seismographic equipment that uses a blast to send out pressure waves and records their echoes from deeply buried rock surfaces is merely a device that permits 'creekology' to be practiced on elevations of ancient land surfaces that are now lost from view. The modern techniques of drilling these buried 'arches', 'domes' or other 'structures' – without understanding the source of the oil and gas – yields a practical find of oil less than once in ten drilling efforts. Thus it would seem to be very important that drillers know how the oil got to be where it is so that they can eliminate at least some of the million-dollars-a-hole wasted costs.

Many independent wildcatters have been victims of this lack of knowledge because they are often driven out of business even though they follow the best advice available. Only the huge oil companies, which by the several devices of tax write-off and of passing on the costs of 10 percent efficiency to the public, have been able to operate this way without penalty. Unfortunately, their ability to survive the ten-to-one failure ratio has served to insulate them from any sense of urgency to pin down the origin of oil. Until 1974, when they began the 'Florida fiasco', the low success ratio hid from them the appalling fact that the 'oil is where you find it' credo also conceals tremendous errors in the governmental estimates of yet-to-be discovered reserves.

It was this laissez-faire attitude that allowed, for example, the assumption that since considerable Gulf of Mexico-fringing real estate was made up of rock layers that often contained pools of oil, it was considered likely that unexplored real estate with similar rock layers would yield a predictable oil-to-square-mile bonanza if enough holes were drilled to find it. In fact, the Gulf of Mexico basin experience – it being the most-drilled basin in the world – was used as the yardstick for estimating the undiscovered reserves of the rest of the United States submarine shelves, and probably the rest of the world's reserves, too.

When ready, the oil consortium stirred up some congressmen and others to resolve the conflicts between state and federal jurisdiction and to open for drilling that vast area of undiscovered petroleum reserves – the continental shelf of peninsular Florida. With the insouciance of well-to-do salesmen spending your money from their expense accounts, the consortium bid millions of dollars for the right to spend other millions of dollars on drilling for oil on the Florida Shelf. The consortium knew exactly the prime locations to bid for, because in the years prior to the bid date, vessels had been sent back and forth over the Shelf gathering the data from which present-day 'creekology computers' could draw maps showing the locations of buried 'structures'. During 1974 and 1975, the drills probed deeply into the Shelf rock layers without success. Finally, in late summer of 1975, complete failure was acknowledged, drilling equipment was withdrawn and operations were suspended indefinitely.

Simultaneously, the United States Geological Survey, the government agency responsible to us and our politicians for advising us about matters geological, announced that its prior estimates of the U.S.' undiscovered oil and gas reserves would now have to be reduced by as much as 80 percent. This, coming at a time when we are already uncertain about the sources of future energy supplies, would seem to be reason for great gloom as we contemplate the economy coming to an unlubricated and chilly halt. Fortunately this is not so for several reasons.

We have vast coal reserves to which we can turn as quickly as required – and solar power or atomic energy reserves after that. But even better is the probability that an absence of oil and gas on the Florida Shelf proves that the undiscovered reserves of the Gulf of Mexico basin have in fact been consistently *underestimated*, and that this underestimation has been applied to similar basins around the world. It all has to do with how oil is formed in the first place, and how the resulting fluid gets into the structures where it is found.

Before each ice age, the Gulf of Mexico looked much as it does now. Then, as now, it was a deformed explosion basin with part of its rim cut out by Cuba and the Yucatan peninsula, which in turn are parts of the rim of the Caribbean explosion basin. As snow and ice were deposited on the continents, sea level fell until the Gulf was cut off from the ocean. Instead of draining completely dry, the Gulf became an inland sea. This sea was still filled with life forms and was fed by the runoff of the growing continental ice sheets, which temporarily supplied additional organic material to the sea.

Eventually, most of the water in the basin froze, enabling this former arm of the ocean to hold a portion of the North American ice sheet. Snow and dusts descended onto these areas of ice, as thickly as they did over land, but within this salty ice a curious change took place.

It is well known to sailors that frozen salt water tends to purify itself. Those who have wintered in the bitter cold of the Antarctic, where the sea water beneath the ice floes becomes supercooled (remains a fluid although colder than the freezing temperature), have long observed a phenomenon known as 'anchor ice'. This ice was named for the fact that it coats the anchor chains and other protuberances of vessels anchored in such waters. Recent expeditions have evaluated this form of ice and found that in the supercooled state salty sea water does not freeze solid; instead, pure fresh-water crystals of ice grow within the solution, and it is these that attach to any fixed object and grow into the larger masses. The growth of such clumps on the shells of small marine creatures or on bottom sediments sometimes reaches a size and buoyancy that brings the clump to the underside of the ice floe. Investigators now believe that it is these clumps, adding ice to the bottoms of the floes, as the tops are eroded or ablated away, that account for the occasional marine creatures found frozen in the surface of the ice.

Another source of information, a U.S. Navy Hydrographic Office publication, advises that entire floes can become fresh water supplies for ships when the floe has aged about two years since any entrapped brine, 'because its density is greater than that of pure ice, tends to settle down through the pure ice. As it does so, the ice gradually freshens . . .'

By this mechanism, the organics and the salt, gypsum and anhydrites that had been dissolved in the water of the basins were condensed and precipitated. Because of this phenomenon, the *Glomar Challenger* drillers found thick deposits of these materials underneath the present floor of the Gulf of Mexico, but being unaware of ice sheet possibilities in this region, identified them as evaporites, thereby implying that they were formed in a hot, dry climate instead.

At the same time that the dissolved materials were settling out of the ice, the dusts and sands from the surface were settling through, either by a similar process or through the action of melting cycles. All of these materials served to thickly cover the organic material on the floor of the basin and to encapsulate it there. Following the encapsulation of the organics, possibly while the ice still lay above them, bacteria began to break down the compounds that made up the bodies of the plants and animals. Working in the absence of free oxygen, the bacteria produced methane gas and, ultimately, the complicated hydrocarbons that comprise crude oil.

All of these steps – death, burial and encapsulation – resulted in the greatest volume of organics being trapped in the deepest parts of each basin. Once formed in this location, the petroleum may take any or all of three courses. (1) It may stay right where it is, in a large pool at the bottom of the basin. (2) Possibly the oil will saturate the rock above or below it. When this occurs, oil shale is produced. (3) The oil and gas may also migrate horizontally, between impermeable rock layers, ultimately reaching the structures of the coastal regions surrounding the basin. There it flows into formations that provide commercially interesting deposits or pools.

If the oil and gas being recovered around the rim of the Gulf of Mexico originated under the deep basin floor and migrated to the rim it should now be possible to prove it.

The search for oil and gas falls into two categories: the broad-scale search for new fields within a basin, and the smaller-scale search for deposits or pools within each field. On the small scale the absence or presence of oil in any given spot, depending as it does on so many variables of topography and structure, yields no proof that this synthesis is more accurate than 'oil is where you find it'. However, on the large scale, minor topographical and structural variations cancel out, and the absence of oil in an entire region calls for a better explanation.

According to the conventional ideas of the formation of oil, the entire coast of the Gulf of Mexico should be prime oil territory. The shoreline has been extensively drilled from the Yucatan peninsula of Mexico to the Florida Keys. Most of this area is highly productive, and new pools continue to be discovered. But for some reason, the Florida and Yucatan peninsulas seem to be almost barren. Between these two peninsulas, however, the Gulf coast is thickly scattered with oil-producing locations. What makes the two peninsulas different from the rest of the Gulf Coast? According to conventional ideas, nothing at all, but hundreds of holes have been drilled in both with negligible results. The conventional hypothesis of oil formation and deposition fails completely to explain the pattern of oil deposits around the Gulf of Mexico Basin.

The synthesis of oil formation presented here holds that when oil is found on the rims of basins, it has migrated there from the deepest parts of those basins. But if there is any topographical feature that blocks the path of that migration, the area on the rim uphill from that blockage will not contain oil. If the oil could flow to Mexico, Texas, Louisiana, Alabama, Mississippi and the panhandle of Florida, why could it not also flow to the Florida and Yucatan peninsulas?

Oil flow in both of these directions is blocked by steep 'scarps'. The Florida fault scarp has been characterized as one of the steepest and straightest escarpments of the entire ocean floor extending for 500 miles along the peninsula from the panhandle to the Keys. Here the bottom drops a mile in a horizontal distance of less than two miles. There are few if any places on land where a plateau comparable to the Florida slope is bounded by such a steep escarpment.

Scarps such as these constitute a barrier to oil migration because they destroy the between-strata 'conduits'. Such conduits are formed by rivers and streams that continue to flow into the basin during the initial stages of an ice age. At that time, buildup of ice in the continental interior regions has lowered sea level, but gulfs and seas continue to receive the reduced flow from rivers. These rivers not only add organics to the gulf, but also are the tools that carve canyons into the shelves and slopes of the basin. As river flow diminishes with the freezing of the water supply, the ice sheet and its included dusts becomes an impermeable lid over the entire scene.

But where the rivers flowed, continuous beds of clean washed gravels and sands have been buried, and these become the conduits through which oil can later flow up from the basin floor.

However, if the permeability of such conduits is destroyed by subsequent earth movements, oil and gas cannot move through them.

The Florida and Yucatan peninsulas are made of Gulf-floor limey muds that have been elevated above sea level and hardened into rock. Had they remained in place after being elevated, doubtless vast quantities of oil and gas would have flowed into their structures. But, as evidenced by the topography, something failed to jell, and they both settled back into the Gulf basin a little bit. That settling formed the escarpments, which in turn show that the buried river conduits have been collapsed or crushed.

For a hypothesis to be of any but academic interest, it must have a valuable predictive capability. What then are the criteria for finding oil under this synthesis?

The setting for oil genesis must be a basin that holds water but can be isolated by falling sea levels, for only in a water-filled basin does organic material become the hydrocarbon of gas and oil. The basin rim must be at an altitude near the ice age frost line, since the basin must hold an ice sheet in order to provide the necessary conditions for the formation of oil.

The center of such a basin would be the prime area for the finding of oil, but if the basin has been reflooded, technical considerations may, at present, require that only the rim be probed.

Recognition that conduits are a fact of oil and gas migration makes it possible to study basin topographic data to great advantage. The 'arch' surrounding such basins is the long-buried rise that marks the rim of the explosion basin. Sometimes it can be easily detected, and sometimes it must be found with seismographic equipment.

The arch has proved to be an excellent prospect for oil fields, although the fields on it are always discontinuous. The discontinuity, in fact, reveals whether or not an unbroken conduit has reached the arch. Where a conduit has not been restricted, the arch has oil fields.

Conduits are sometimes easily discerned from marine sea floor charts. The conduits lie under the submarine valleys and canyons that cut the basin slopes from floor to rim.

With these facts in mind, we can examine the other coasts of the United States and evaluate them as potential oil producers. The Arctic Ocean Basin has recently proved to be another great oil supply. Even a cursory study of marine charts of this basin shows that the oil recovery is taking place where the basin arch intersects the submarine valleys. The California coast supports the synthesis. That coast line has long troughlike basins, parallel to the present coastline, that were formed by the ice sheets of earlier ice ages. Here, too, where contours have been relatively undisturbed by recent ice sheet glaciers, may be seen the pattern of submarine valley intersections with coastwise arches and faults that provide the rich fields.

The Atlantic coast, however, presents a different appearance. There may have been a few small basins; for example, the submerged Blake Plateau – underwater off

Jacksonville, Florida – may conceal a small basin in which some organics may have been trapped. But on the larger scale, there is little reason to expect that there was any great oil genesis in this area. Ice-age lowering of sea level did not cause any large bodies of water to be isolated on or near the Atlantic Coast. The water of this area simply ran off into the unfrozen basin that lies under the Sargasso Sea now.

For these reasons, it is the prediction of this synthesis that no substantial oil recovery will be made off the Atlantic coast of the United States.

The prediction need not be cause for great concern, however. The absence of oil in the Florida and Yucatan peninsulas, and its presence in the rest of the Gulf Coast, indicate that basins such as the Gulf are larger repositories of oil than has been previously estimated. Indeed, it now appears that the oil and gas being recovered from the rich continental shelf and coastal plain oil fields of Mexico, Texas, Louisiana, Mississippi, Alabama and the panhandle of Florida may be only the seepage coming from vast deposits on the floor of the Gulf. They are now pumping only a small fraction of that oil – that which has migrated from the bottom of the Gulf to its rim. At the bottom of the basin may be an oil field whose volume surpasses any that has been tapped before. Similarly, as we are now discovering in the North Sea, rich fields may lie under the floors of other flooded basins, such as the Caribbean Sea and the Arctic Ocean. When drilling techniques are developed which can reach these areas, and the techniques are under development today, we may have vast stores of oil to serve our future needs.

At present, the oil industry – which is very secretive about its techniques – is believed to employ only two criteria in its exploration for oil: the location of 'structures' by the use of seismographic, gravimetric and magnetic surveys and the premise that the area to be surveyed must be a closed or 'evaporite' basin.

Instead, if they use six factors in their search, the

remaining undiscovered oil in the world should be readily located. (1) In general, since oil-producing organics flourish during the height of water ages, basins within or closely surrounded by large land areas above present sea level have the highest probability of containing oil. (2) The basin must be shaped so that its water does completely escape when sea levels are drastically lowered. (3) That closure must be near present sea level so that an ice sheet may cover the basin during ice ages. (4) The topography of the slopes and floor of the basin must be evaluated to find the valleys which represent buried conduits. (5) The routes of the conduits must be traced to ensure that waterfalls, scarps or other hazards have not closed them. (6) Structures must be found that lie along or in contact with the conduits, since these are the collection points that make drilling profitable.

Phosphate Rock

Ice sheets and ice ages have provided us with other beneficial minerals. The configuration of coal and phosphate deposits suggests that they were also formed under conditions that no longer exist in the world. Let us re-examine these deposits in the light of the conditions imposed by a worldwide ice age.

Phosphate deposits always contain huge deposits of the fossilized bones and teeth of long-dead animals. Therefore it is reasonable to assume that it is the phosphorus content of these bones that has been leached out to form the 'phosphate rock'. However, the deaths of large numbers of animals within a limited area is not, by itself, sufficient to form phosphate deposits. The slaughter of the buffalo in the American West during the last century left millions of skeletons to bleach in the sun, but no phosphate deposits developed as a result of this carnage.

The huge deposits of ancient bones that make up commercial phosphate rock deposits are staggering, even to those who acknowledge the catastrophes that Earth has suffered. Perhaps ice ages bring about these deposits by causing unusual herding of animals. Perhaps, too, the animals were first killed by the onslaught of the ice age and only later were piled up in vast deposits by streams and surf lines.

Regardless of how the bones were accumulated, the chemical conversion of those bones into phosphate rock and fossils suggests that they were deposited under conditions that do not exist today. In his book, *The Formation of Mineral Deposits*, Alan M. Bateman suggests that these deposits were formed under very unusual marine conditions; that no sedimentation took place as they were laid down; that the deposits were cut off both from sea and air; that cool temperatures are indicated; and that there is a surprising amount of fluorine present.

The largest volume of phosphate being mined in the United States today comes from the Bone Valley formation in central Florida. Such a combination of conditions in Florida today is, indeed, very unusual. But let us consider the phosphate deposits as they were laid down during an ice age. The freezing temperatures on the continents can account for the apparent cool conditions. Burial by snows and volcanic dusts would provide the necessary insulation from air that is shown by the preservation of the hydrocarbons and iron sulfide, and the chemically charged volcanic dusts would also provide the fluorine that is found in such high concentration.

Another good indication that phosphate rock in central Florida, and presumably elsewhere, represents a staggering quantity of dead animals from which the phosphor of bones and teeth was leached may be found in the peculiar accumulation of radioactivity associated with phosphate rock deposits.

Radioactivity is everywhere about us, but animals and plants seem to be natural concentration mechanisms for it. A recent publication of the American Nuclear Society advises that although the federal standard for maximum allowable concentrations for human consumption in drinking water is two picocuries per liter, sea water in the Gulf of

Mexico contains 350; scotch whisky, 1200; and cow's milk, 1400.

It is apparent from these figures that while runoff of water from the land tends slightly to concentrate radioactivity in the Gulf, it is the grains that go into making whisky – and the eating habits of grazing animals – that are many times more efficient concentrators. It follows then that the dangerously high levels of radioactivity found in phosphate show the direct relationship to animal origin. It is also understandable that both oil and coal have relatively high radioactivity, since they, too, are animal and plant remains. Finally, the fact that each wave of life-form extinction, time after time in the dim past, is directly related to, and accompanied by, the formation of oil, phosphate and coal beds – all at the same time – leaves little room for doubt about the animal origin of phosphate.

The animals involved in the Florida phosphate beds were not microorganisms. They were the ancestors of the elephants, the camels, the horses, the hippopotami, the pigs, the tigers – literally the entire panoply of the life forms with which humans are associated.

Even humans may have contributed their forms to this vast supply of chemicals.

The Good Earth

The great continental ice sheets may have bestowed another blessing on the world in the form of soils. Geology texts state that soils are derived from the decomposition of the bedrock that underlies them. This view conforms with the doctrine of uniformity and in a way is a very comforting thought. If soils are derived from underlying bedrock, then we need not be too concerned about conserving them. Decomposition of bedrock will continually supply new soil to support plants, animals and us.

But look at some of the conditions under which we find soils. In vast areas in and near the Arctic Circle, fields of ice left over from the Pleistocene ice age are covered with soils and plants. Were these soils derived from the ice beneath them? Clearly they were not. In many places, the soils which lie on top of decomposing bedrock also have little in common with the rocks below them. In fact, some soils lie on rock which is not decomposing at all. We have all seen this phenomenon in roadside cuts through hills and mountains; a sure sign that at least some soils are deposited in a far different manner than is commonly supposed.

When rock such as granite or basalt is ground to a fine powder, it results in an end product that is called 'rock flour' and this is commonly found where glaciers have pulverized the materials over which they moved. This rock flour has a lack of fertility that one might expect of a chemically inert material, and it does not encourage the growth of vegetation.

On the other hand, it is well known that those who live on the flanks of active volcanoes are in constant danger from eruptions, but because the ash and muds which flow from volcanoes are enormously fertile, people in many nations constantly risk the hazards of eruptions in order to reap the rich bounty of crops that will readily grow in this kind of ground-up rock. So it seems unlikely that it is the slow, particle-by-particle deterioration of rock that makes the world's thick and fertile soils. It is likely, however, that the type of catastrophic vulcanism associated with ice ages ejects great volumes of chemically active dusts into the atmosphere, and that these dusts spread worldwide over the ice sheets.

Consider, therefore, the probability that the soils which now cover the highlands of the Earth were originally accumulated in the ice sheets as they grew. The ice sheet of Greenland is white and clean looking, but that is only true of its surface. In fact, the continental ice sheets of the ice ages were heavily loaded with volcanic ash, dusts and sands and with silt blown up on them from the newly dried ocean floors. In time, these dusts and sands may even have made a soil cover on the surface of the ice sheets that was thick enough to support life forms as the ice sheet melted. This would be a likely place for post-ice-age-peak ground cover to begin, since the plants would only have to send their roots down close to the ice in order to be assured of a constant source of moisture.

One of the most interesting, and most puzzling, types of soil is loess. It is a deposit of loosely arranged, angular grains of calcareous silt loam, typically intermediate in fineness between sand and clay, and of remarkably uniform mechanical composition. Normally it is without stratification and breaks off in vertical slabs, with the result that perpendicular cliffs are formed. But despite the fact that loess can form a vertical cliff, it is a soil and not a rock. The particles which make it up are separate and distinct, not fused together in any way. The remarkable properties of loess stem from the fact that the individual particles which make it up look, under magnification, as if they have been carefully assembled to form a material of maximum strength. It is as if each grain were put into place, one at a time, and lined up with the others on purpose. An analogy to this remarkable material can be made with common bricks. If truckload after truckload of bricks are tossed into a pile, that pile will be a broad mound with sloping sides. Bricks have a natural angle of repose just as does sand. But if the bricks are carefully stacked, one on top of another, vertical walls can be built that are stable, even without mortar. It is the careful stacking of loess particles that allows them, too, to hold a vertical surface. Once loess has been disturbed, it loses that structure and can never again maintain a vertical surface.

Loess deposits are without stratification, even though deposits in China are as much as a thousand feet thick. This suggests that neither wind nor water brought this remarkable material to its final resting place. Had the loess been borne by either of these means, a process called 'sorting' would have caused the heaviest particles to settle first, followed by lighter and lighter particles, in distinct layers. This did not occur.

Loess is found evenly distributed over the areas that it

covers. Both hill and valley have an equally thick cover of it. Some suggest that loess was deposited by blowing wind, but this could not account for the odd deposition. Blowing wind would tend to erode the loess from the hilltops and deposit it in the valleys, but the uniform thickness of the deposits belies this idea. It has been suggested that the dust and sands were carried by a wind storm which suddenly stopped, allowing these materials to drop straight down onto both hill and valley evenly. This might have happened, but then air sorting would have taken place and the deposits would be stratified.

It is clear that loess was originally windblown dust, but the method by which it is deposited as it stands today needs to be adequately explained. Loess does not appear to be derived from the rock which underlies it, so it is reasonable to assume that the wind carried it to its present position. Also, the presence of microscopic seashells in loess suggests that some of the dusts and sands originated on distant ocean floors.

These dusts probably were deposited on and in the slowly growing continental ice sheets. Later, as the ice slowly melted, particles were 'stacked' into their distinctive arrangements, and finally the loess and soil that covered the ice sheets developed layers of vegetation that prevented erosion. As the ice sheets melted from beneath these new layers of soil, the vegetation, soils and loesses were lowered as uniform blankets over the lands that they covered.

The primary reason that loess, soils, phosphate and oil have not been previously identified as ice sheet effects by those responsible for geological analysis is apparent. The uniformitarian absolutists in charge have decreed that catastrophes shall not be recognized and that ice sheets consequently cannot exist near Earth's Equator. Geologists have therefore been forced to misconstrue evidence that lay in the ground all about them.

Coal

The formation of coal also has mysteries that may now be solved. According to the uniformitarians, the process of coal formation begins with an accumulation of plant remains in a swamp. This accumulation is known as peat, a soft, spongy, brownish deposit in which plant structures are easily recognizable. Supposedly, great eons of time, coupled with *the pressure produced by deep burial*, gradually transforms the organic material into coal.

But consider the locations where coal is found today. As in the rest of the world, the largest coal fields in the United States are found in its 'craton', which is the geologic term for the nonmountainous interior or 'plate' of the continent. These fields are identified as the Appalachian, Illinois and Mid-Continent, and all are characterized as being in regions where the strata lie flat. The Mid-Continent field includes the coal fields of Missouri, Iowa, Kansas, Oklahoma and northern Texas. Over these extensive areas, the topography is flat and the coal so near the surface that it is mined by stripping.

Strip mining is a technique used only where a relatively thin cover of vegetation, plus soil and soft rock, may be removed by bulldozer and dragline. But the description of the process by which peat is converted to coal specifies that 'pressure produced by deep burial' is necessary for the conversion. Burial by what, and when, and how deep?

The textbook informs that it has been established that the formation of bituminous coal requires the pressure of burial under the equivalent of at least 7500 feet of rock.

The uniformitarians must therefore suppose that vast areas of the central United States were once covered by rock layers one and one-half miles thick, which have since mysteriously disappeared. Furthermore, the layers of coal occur in cyclothems – recurring cycles of strata indicating that the same pressure was applied, removed and reapplied many times.

The existence of coal must therefore amaze them, since

the coal beds have formed on the surface of a flat area that maintained that condition with no trace of distortion while a mile-and-a-half thickness of rock, covering millions of square miles, came and went many times – leaving no trace of its identity!

Let us consider another possibility. There is a material that could thicken over these beds, exert enormous pressures over vast areas and then disappear completely. The material that can do that is layers of ice, and the continental ice sheets that have covered the United States many times fit the picture exactly. A four-mile thick ice sheet exerts the same pressure on an area as would 7500 feet of rock, and the ice sheets had to average a thickness of six miles over all the continents.

The formation of anthracite coal, rarer than bituminous, requires even greater pressures. Peat must be buried under the equivalent of at least 19,000 feet of rock in order to be turned into anthracite. It, too, is mined near the surface from mountains that do not exceed about 5000 feet in height. But again, where ice sheets average six miles there is little doubt that thicknesses up to 10 miles occur where conditions are appropriate, and ice of that thickness exerts pressure adequate to convert peat into anthracite.

So, in a strange way, coal gives us a permanent record of the maximum thickness the ice stood in a given area, by whether the peat was converted into bituminous or anthracite.

The ice ages were enormous catastrophes that wiped out whole species. But at the same time, the ice ages also laid down new minerals and fertile soils. Those who believe in uniformity must also believe that somewhere somehow, nature is replenishing the supply. Their faith is touching; their teaching is dangerous.

9

Mountain Building

Continents Do Not Drift

Charles Lyell in his book *Principles of Geology* proposed that the Earth, over the eons of its existence, changed but little, even though there were dynamic forces in effect which made temporary minor changes in the Earth's crust. Lyell saw the changes in sea and land as cyclical: erosion caused the reduction of the land; wind and water bit by bit chipped away both mountain and lowland, washing the detritus out to sea. After the land had been reduced and the sea floor covered with the product of that reduction, the sea floor would slowly rise so that the whole process could begin again.

This idea eventually developed into the hypothesis of isostasy, which says that both ocean floor and continent float on the molten mantle of the Earth, just as blocks of wood float in water. The Earth's surface was thought to be entirely made up of huge sheets of cool surface material of different densities. The denser material sank lower into the liquid mantle, and thus became sea floor. The lighter blocks of material made up the masses of the continents, since they floated up above the denser sea floor.

It was proposed that this relationship of the materials of different densities was predominantly stable, but that it would ultimately be altered by erosion. Erosion of the land caused the continental area to become lighter, and the deposition of that material caused the sea floor to become heavier. This caused an imbalance in these floating bodies. The decreased mass of the continental areas caused them to be buoyed up, and the increased weight of the sea floor caused it to sink deeper into the molten rock upon which it floated.

This explanation of isostasy was later augmented by the idea that the sedimentary material that was deposited on the sea bottom ultimately 'broke through' into the molten rock of the mantle as deposits got deeper and deeper. These sediments, metamorphosed and expanded by the heat, supposedly floated up eventually to form mountains.

Isostasy has fallen out of favor as an explanation of the formation of the features of the Earth. But the idea that the continents floated on the Earth's mantle of molten rock spawned another hypothesis, that of continental drift or 'plate tectonics'. In 1912, Wegener published his proposal of continental drift. He supposed that the dry land of the Earth was once a single, vast continent that he named Panagea (from the Greek for 'all' and 'Earth'). He argued that this primeval continent began to split apart toward the end of the Mesozoic Era. The fragments drifted across the Earth's surface and by the Pleistocene Epoch were in position as our modern continents. Those who advocate the hypothesis of drifting continents estimate a maximum drift rate of about six inches per year and account for the thousands of miles between continents by suggesting that the journeys of the continents to their present locations took many millions of years. They also suggest that mountains are formed as a result of rock in motion pushing against the edges of these continental plates and wrinkling or folding them. The force that powers all of this drifting and folding is given as the convection of molten material within the mantle of the Earth. Examine the evidence and conclusions of the adherents of this hypothesis carefully.

Much has been made of the similarity of the composition of mountains of eastern South America and western Africa to show that these continents were once joined. Geochronologic studies of Precambrian igneous and metamorphic rocks in western Africa and the eastern bulge of South America, and radiometric age determinations of many samples show distinct belts of roughly similar age on the two continents. 'Drifters' therefore propose that rocks of similar age in Africa were once continuous with rocks of the same age in South America.

Evidence of the similar ages of the rocks of the two continents does not warrant the conclusion that they are derived from the same source. A photograph made by an orbiting astronaut-geologist was captioned, 'Morocco's Cap Juby is near the lower center here. Light spots near it are salt flats ... Streaks of cirrus in the upper left are over the Atlas Mountains. At the right is the Hamada du Dra, a plateau underlain by the Tindouf syncline. Discordant geologic structures on each side of the Atlantic are often cited to support the theory of continental drift, but this photo of Morocco and Spanish Sahara shows concordance to the African shore.' In other words, the coast of Africa is rimmed by continuous mountain ranges, as is South America. We may therefore safely conclude only that the forces which shaped these two widely separated mountain ranges were similar, and that they occurred simultaneously in both South America and Africa.

If it were true, however, that the continents drifted apart over millions of years, that journey would be very difficult. Consider the characteristics of the rocks supposedly involved in this drifting process. The strength of basalt is of the same order of magnitude as that of granite. The average crushing strength of 41 samples of granite is 1500 kilograms per square centimeter. The drifters suppose that the ocean floors are made of basalt. Therefore, to move, the continents must break this rock or be broken by it.

The supporters of the theory of continental drift believe that the continents are made up of plates that are sufficiently plastic that they can be folded into mountains. According to this idea, all rock has properties somewhat similar to 'Silly Putty'. While hammer blows will shatter it, if pressure is applied for millions of years, rock can be bent and folded into mountains. Clearly, it can be proved by experiment that rock does not behave like Silly Putty, so this is an absurd proposition. Similar thinking on the part of the adherents of the theory of isostasy has led to statements 'that no rocks are strong enough to support the weight of even comparatively low hills', much less mountains. New thinking is needed, since it is obvious that rocks do not bend, and that they do support mountains.

The continents, however formed, are subject to the ravages of time. The material being eroded from them which is not trapped in interior basins is continually being deposited around them and is not carried toward the center of the ocean basins in any great volume. Therefore, the sediments of the continental slopes are the overflow of the shelves, piling up at the angle of repose. The continents are continually spreading by enlarging this sedimentary border. The 'angle of repose' of a material is the maximum angle to which a material may be piled without collapsing. This principle is easily demonstrated with beach sand. If dry beach sand is piled up, the mound grows in height until it collapses, leaving a broad, low cone. Continuing to pile more sand on this mound will make it grow larger, but the angle made by its sides will never get any steeper. The angle thus formed is called the 'angle of repose' of sand.

The angle of repose of materials deposited in water is ordinarily steeper, but when we examine the underwater slopes just off the edges of the continents, we find that they are gentle enough so that the family car could be driven up them in high gear without strain. Clearly, those slopes were gentled by something other than sedimentation.

Each falling sea level produced a retreating surf line which reworked the slope as it passed across it. The later melting of those ice sheets provided another moving surf line as the sea level rose. Successive ice ages and sea level changes ultimately planed these slopes to the very gentle grades which they now hold.

Presuming, however, that mountains parallel to the sea coasts were the results of drifting continents, it is reasonable to assume that the sediments of the sea floor as well as those of the continent would be under extreme compression at the junction of the two. Pursuing this line of reasoning, then, we would expect to see wrinkling of the sea floor where it lies next to mountains that were supposedly folded up by the collision of sea floor and continental plate. We see no such wrinkling, however. Those supporters of the theory of continental drift can point only to short submarine trenches which fringe the sea floors in a few places in the world. There, they suggest, 'subduction' of the crust pulls down the sediments into the Earth. Even there, these so-called trench slopes are also very gentle.

The forces that supposedly cause the continents to wander about the Earth are heat and convection. Convection is a movement within a fluid that occurs as a result of its localized expansion and rise. Drifters speculate that when the Earth's mantle material is heated and begins to expand, it moves upward for about 25 million years, gradually 'speeding up' until it attains the rate of five inches a year. There is a serious flaw in this concept, however, because it fails to take into account the fact that when convection begins in a sphere such as Earth, the heated material must fill an ever-increasing conical volume as it rises. As Dr A. A. Meyerhoff points out, such expansion would cause the material to immediately cool itself. In order for further movement to take place, more heat would have to be applied, and the source of that heat would have to move with the heated mass.

Another problem is that convection currents and thermal processes must have been continuous to have supplied a steady or only slowly changing amount of energy throughout geologic time. Yet one of the peculiar features of mountain formation is that it is not at all uniform in time. Periods of mountain building are definitely not continuous.

Recent seismic research has also developed evidence that proves that convection currents do not exist within the Earth. As a result of the need to monitor nuclear test explosions, worldwide, very sophisticated sensing devices have been developed. These seismometers enable us to 'see' the composition of the Earth's interior by observing how pressure waves move through the Earth. In 1973, reports were published showing that sharp variations in rock properties exist in the top few hundred kilometers of the Earth. That is just the region where, according to the new plate-tectonic view of the Earth's crust, there must be convection currents to account for the movements of continents and crustal plates. But, instead, the seismic research shows that the Earth is composed of many distinct layers of flat-lying material. If there were convection currents, they would show up like the swirls in a marble cake.

The proposals of isostasy and of drifting continents both developed from the assumption that the Earth is a molten ball upon which a thin crust of cooled material floats. This very basic assumption has now come under fire as the result of satellite observations. The Earth does not respond to distortion like a ball of molten liquid rock, but rather is quite rigid. In places it is nonspherical – ranging from +62 to —79 meters from a perfect sphere. And since the Earth has mascons (mass concentrations) and craters, it cannot be considered to be fluid.

One of the fundamental assumptions of the theory of drifting continents is the assumption that most sea floors are spreading. This idea is supposedly supported by the evidence of magnetic anomalies on the sea floor. Supposedly, as the sea floor spreads from a mid-oceanic ridge, new molten rock rises to the surface and cools. In doing so, the newly cooled rock takes on the magnetic orientation of the Earth. Later the rock moves outward away from the mid-oceanic ridge, forced to move by the upwelling of other new rock. If this movement were to occur continuously as the Earth's magnetic field reversed itself several times, symmetrical bands of oppositely magnetized rock would lie on either side of a mid-oceanic ridge.

However, the developers of this theory were not confident that this was the only explanation of the magnetic anomalies on the sea floor. They acknowledged that the same magnetic patterns could be caused by complicated electrical currents within the Earth.

There is yet another explanation for the magnetic anomalies on the sea floor. They may be erosion exposures of sediments that are each alternatively magnetized as they come into the influence of the preceding layer. The magnetism of rocks is an imperfectly understood phenomenon.

Both the physical characteristics and the interpretation of the magnetic anomalies have come under heavy fire. Meyerhoff noted that although magnetic profiles from the oceans were fairly abundant by 1960, symmetry was not even considered until the Fred J. Vine and D. H. Matthews paper was published, and then symmetry was instantly assumed to be the rule; but recently the fact that wide zones of magnetic stripes 'dive' beneath continental margins in many places has become apparent. This is not at all in accord with drifting continent hypotheses.

Another aspect of the magnetic field of the Earth is used to support the theory of drifting continents: the study of paleomagnetism. Drifters support their hypothesis by noting that subsequent lava flows indicate different magnetic orientations in different continents at the same time. Their proposal is based on the assumption that the Earth's magnetic field has always approximated the Earth's axis of rotation.

Theoretically, if lava flows were to take place all over the Earth at the same time, when those lava flows cooled they would all show magnetic orientations that coincide with the Earth's present magnetic field. If one were to draw lines around the world from each lava flow showing its magnetic orientation, all those lines would coincide at the north and south magnetic poles of the Earth.

But lava flows created at the 'same time' in the ancient past have magnetic orientations that do not coincide with each other today. Drifters therefore assume that since the lavas were formed, the continents on which they were formed have been twisted and moved about the Earth so that their magnetic orientations have been scrambled. Thus they conclude that the inconsistent nature of the magnetic orientations of ancient lava flows supports the theory of drifting continents.

But look at the assumptions upon which this conclusion is based. One assumption is that the Earth's geographic poles must have coincided in the past with the Earth's geomagnetic poles. Clearly that is not so in the present; why should we think that it was true in the past?

Indeed, evidence from our own age indicates that any conclusion based on the notion that the Earth has a stable magnetic field is highly suspect. It has long been known that magnetic declination changes. Not only are there slow changes in declination, but there are also changes in inclination and intensity. Records from Gallo-Roman time to the present indicate great changes in the declination and inclination at Paris. But records for the entire Earth show that those changes in the magnetic field of Paris are regional rather than global.

Between the years 25 and 1740 A.D., the magnetic inclination at Paris changed by 12 degrees and the declination changed by about 30 degrees. If lava flows laid down during each of those two years were checked for magnetic orientation they would be substantially different. Using the logic that concludes that similar magnetic differences confirm the theory of drifting continents, one would have to conclude that Paris drifted a considerable distance between the years 25 and 1740.

This, of course, is an absurd conclusion. But it does point out the fallacy in the logic of this argument for drifting continents. Present-day observations show that the Earth's magnetic field alters substantially during periods measured in hundreds of years. In geological terms, this variation is one of amazing speed. And in long-term measurements of time, being able to identify a date within a million years is excellent accuracy. Many wild gyrations of the Earth's magnetic field are possible within a million-year period. For this reason, any magnetic inconsistencies in ancient lava flows can be completely explained by magnetic oscillations that take place so fast that current dating systems are too inaccurate to spot them at all.

The proposals of isostasy and drifting continents both are shown to have serious drawbacks when scrutinized closely. The 'evidence' of drifting continents primarily consists of the fit of Africa and South America and the presence of magnetic anomalies along mid-oceanic ridges. But similar magnetic anomalies exist other than along such ridges, and there is serious doubt that the findings have been properly interpreted.

The believers in continental drift say that, over millions of years, rock may be folded and deformed into mountains. But over those same millions of years, the shapes of South America and Africa have stayed rigid enough to still fit together perfectly. Mountains are supposedly folded up by sea floor material which is continually being shoved under the land, but that ocean floor itself shows no deformation whatever.

The theory of continental drift had been around for a long time, but it didn't catch fire until recently. Prior to World War II, most of the supporters of the theory of continental drift lived in the Southern Hemisphere, particularly in South Africa. Geologists and geophysicists of the United States and Western Europe, felt that there was little or no support for the motion of continents or the wandering of the poles. In fact, it was hardly respectable to entertain the possibility of such movement.

Close examination of the structures of many mountain ranges also refutes this belief. The ranges which have been raised over flat-lying bedrock pose a difficult problem for the drifters. Here, miles of flat-lying sediments have been horizontally compressed into mountains; but the basement rock over which they slid shows no compression distortion to match the surface effect.

If the horizontal force came from drifting continents, it must of necessity have been exerted miles under the surface and its effect should be greatest there and diminishing above it. Instead, the surface has been crumpled and folded over a smooth bedrock as a tablecloth wrinkles when it is pushed across the surface of the tabletop. Like the tablecloth, if these mountain ranges were pulled out smooth, they would again be flat layers of hardened sea-floor muds hundreds of miles across. The continental drift premise simply doesn't provide the answer to this movement. Other proposals such as the effect of earthquakes or of sliding downhill on tilted beds that later leveled have been competently disproved.

The refutation of the hypotheses of isostasy and plate tectonics reopens many questions. Primary among them is the question, if the mountains were not raised by drifting continents, what force did raise them?

The Footprints of the Ice Sheets

The great continental ice sheets that covered all the present land area of the world from pole to pole were, for the greater part, stationary bodies. The ice sheets left traces of moving glaciers at their edges, but stood still above the land over most of their area. It is absurd, however, to think that an ice sheet up to 10 miles thick could rest on an area without affecting it in any way. In fact, the land was much affected, but in ways that have not been previously recognized.

The technical name for mountain range building is orogeny, and the forces required to form mountain chains may be divided into two components: vertical forces and horizontal forces. Some uplifting is caused by volcanic activity in which lava flows under existing strata or onto the surface. Uniformitarians have held that the rest of the horizontal and vertical forces of mountain range formation are caused by isostasy or drifting continents.

But consider instead the effect of a weight placed upon a soft, plastic material. When a person steps in soft mud, his weight forces mud from under his foot; he sinks in. The mud then flows from the area of high pressure directly under the foot to an area of lesser pressure that surrounds the foot. The area of the footprint itself is depressed below the surrounding mud, and the mud which formerly occupied that space is displaced to form a ridge around the footprint. That ridge is not only higher than the bottom of the footprint; it is actually higher than the surrounding material.

If the footprint were made in fairly thick mud, it would retain its shape even after the person who made it walked on, even though it might slump slightly after the pressure of the foot was removed. If that mud were later to turn into rock, a permanent cast of that footprint would remain, including its central depression and raised rim.

It was in this manner that tremendous, inexorable forces of gravity and giant, standing ice sheets supplied the vertical and horizontal forces that squeezed up mountains on the edge of the continents. Prior to the coming of each ice age, some areas of the continents contained shallow seas that built up layers of mud on their floors and margins. These muds may have had limey, sandy or clayey characteristics, but all had the quality of being able to later harden into rock. Underneath these layers of soft mud lay hard rock strata, made flexible by widespread fractures within them. Still deeper were layers or pockets of molten rock or magma. This magma, like the muds, was soft and able to flow. Under the great pressure of the miles of ice, the soft sediments and magmas tended to flow outward, away from the center of the ice sheets. Just like the mud under the weight of a man's foot, the soft materials frequently found relief from pressure by flowing into other sediments at the edges of the continents. There they piled up as ridges.

Later the magmas that lay under the newly elevated mountains hardened into rock. Examination of the mountains of the world shows just such structures, called batholiths by geologists. These batholiths are also called the 'roots of mountains'. We would expect these batholiths to resemble the ridge surrounding a footprint in mud, and they conform to this expectation.

The structure of batholiths clearly implies that they are created by ice sheet pressures. Vulcanism-generated intrusions of magma into sedimentary rock assume shallow, horizontal forms, such as laccoliths, sills and lopoliths, which are flat formations that do little to elevate the sedimentaries they intrude. Very rare are the vertical formations such as the cores of volcanic vents or dikes which must be supported during their cooling by previously hardened rocks.

Batholiths, on the other hand, are remarkable for their elevation above their surrounding territory. They extend from deep beneath the surface of the Earth to altitudes in the thousands of feet carrying masses of sediments above them. Clearly, batholiths are not merely large versions of the other types of intrusions. Their great height, compared to the size of their bases, sets them apart.

The elevation of the batholiths implies that some condition forces the magma intrusion to become a long, high ridge rather than a flat lens. This condition is satisfied by the continental ice sheets. Soft materials underlying an ice sheet are elevated in its voids and margins. There they cool and harden while held in place by the ice sheet. Ice sheets certainly did have the enormous weight needed to elevate and hold these materials.

The three-mile fall in sea level formed ice sheets averaging six miles thick on the continents. Prevailing oceanic winds dropped much of their water as snow on the former coastal areas, forming thicker ice sheets in these areas, as much as ten miles thick. This ten-mile thickness of ice has sufficient weight to elevate rock, sediments and magmas *three miles* above the surrounding terrain.

According to uniformitarians, there should be no limit to the elevation of mountain ranges, since the proposed convection currents in the magma that are capable of moving continents should be adequate to the task. Therefore, we should expect to see mountains five, ten or even fifteen miles high. Such mountains would be very useful since an 80,000 foot mountain would permit observatories to be built on Earth, instead of having to be on a satellite, to get above Earth's atmospheric climate zone which so greatly diminishes clear observation of the heavens.

But nonvolcanic mountains on Earth do not extend above the climate zone. In fact, they seem to find a height limit at about three miles above the source of pressure for their formation, and considerably below the climate zone upper limits.

This fact suggests the following. Raising such materials three miles with a counterweight made entirely of ice would require that the ice be approximately three times as high – about ten miles. And the upper limit to which ice can be formed in today's atmosphere would be the upper limits of the climate zone. At present, this is the vicinity of jet streams and very high clouds – about thirteen miles above sea level. Mt Everest and K-2 in the Himalayas are about six miles above sea level, but are only two miles above the western China plateau that contributes its mass as a counterweight. Therefore, to raise them to their present heights, ice would have had to form three times two – or six – miles over the plateau; a total of ten miles above sea level. This it can easily do.

The same correlation is found with high mountains throughout the world. The agreement between mountain height limits – the ice sheet height limits – and climate zone height limits provides another measurable factor in evaluating the size of now-missing ice sheets.

To further evaluate the validity of this hypothesis of mountain building, consider other properties of mountain ranges. First, mountains occur in long, narrow belts. Since ice-sheet-formed mountain ranges would primarily be raised ridges along the margins of ice sheets, they should be shaped this way.

Second, the sedimentary rocks exposed in the chains are largely those which had earlier been beneath the sea. The thick masses of soft sediments that rim the continents or sometimes make up the floors of continental seas were overloaded by ice and squeezed up to form mountains. These mountains would, of course, be made of the materials previously deposited on the sea floors.

Third, materials originally deposited nearly horizontally

in the sea have been compressed as a part of the mountain building cycle. If the mountain ranges were flattened out, they would expand horizontally by tens or hundreds of miles. This, too, is compatible with the theory of ice-sheet-formed mountains. The soft sediments under the ice sheets were sometimes squeezed like toothpaste from a tube. The squeezed-up material piled up on the edges of the ice sheets in tremendous wrinkles. Since this was a dynamic process, sediments were periodically extruded from under the ice sheets and over their edges, and more and more material was squeezed into the same area. If the folds and wrinkles caused by this process were flattened out, they would indeed be many times wider than the mountains they make up.

Fourth, since mountains are long, narrow structures, the force that raised them must have acted largely perpendicular to their axes. The force which folded the sea floors into mountains was applied to them along a broad front. Adherents of the drifting continents theory believe that the force comes from outside a continent, rather than its interior. Force and folding can occur along the same line from either direction, but in order to form a *curved* mountain range (as most are), an exterior force would have to simultaneously come from many different and separate points. However, simultaneous force from the interior of the curve – and the continent – is always exerted from the highest part of an ice sheet.

Finally, the fact that the margins of the continents are the primary scenes of orogeny has been emphasized by geology textbooks. It is apparent that during the ice ages the margins of the continents, which are also the margins of their covering ice sheets, would be the locale for the newest phase of mountain building.

Imagine the Gulf of Mexico and the Caribbean to be full of ice. Visualize that ice sloping up from more than 600 feet above present sea level on the Bahama Shelf to more than 900 feet above present sea level west of the Florida peninsula. Furthermore, move westward with the mind's eye over the frozen expanse of the Gulf of Mexico Basin to find that the shores of Mexico are buried under more than ten miles of ice! To the north and south, both American continents are similarly buried under this vast sheet of ice.

These ice-sheet-thickness minimums are easy to establish because ice is about one-third the weight of mud and rock. So, since the limey muds of the Bahama Shelf were pushed up into ridges that hardened into rock which is now standing more than 200 feet above present sea level (Cat Island), it is apparent that the ice lying west of it had to be more than 600 feet thick. (No wonder such great chasms as Tongue of the Ocean and Exuma Sound were washed out of the Bahama Shelf when the ice melted!)

And since peninsular Florida boasts as its highest elevation a 'mountain' 325 feet high, ice in the eastern Gulf Basin had to be more than 900 feet above present sea level.

But it is the western part of the Gulf Basin and both American continents that obviously received the greatest amount of snowfall, probably because the winds rise and lose their snow crystals as they flow eastward out of the Pacific Basin and over the continents. One can almost see these snowfalls piling ever higher in the west, obviously reaching heights of more than ten miles to push up the muds of Mexico and the rest of the Americas into the Western Cordilleras, which stretch from the Brooks Range in Alaska through the Rockies of North America and the Central American Mountains, to the magnificent Andes chain in South America.

It was these western snowfalls, wedging their alternate expansion cycles against the vast sheets of ice that were firmly grounded on consolidated rock in the east, that forced the muds and magmas in the west to elevate the great blocks of three-mile-high mountains now bordering the western sides of the continents.

Evidence that the isthmus of Panama was elevated between formerly unconnected South America and Central America was reported in 1971 by Dr Tjeerd H. van Andel and G. Ross Heath. The two oceanographers from Oregon State University reported that deep sea drilling gave evidence that the isthmus of Panama is an area of elevated deep ocean floor that was forced to rise above sea level. Somewhat later the southeastern part of Costa Rica was also elevated. The island of Cuba is simply another part of the ridge that divides the Gulf of Mexico from the northwestern basin of the Caribbean. The Yucatan peninsula, western Cuba and the underwater ridge that connects them are all elevated sea floor sediments, just like the mountain ranges on the continents.

The ice of the central basin of the Caribbean squeezed up the isthmus of Panama to its west, and with the ice sheet of the eastern basin, also squeezed up the mountains that became the Greater and Lesser Antilles. What we see today as islands are peaks of a great mountain chain, formed by ice sheets and partially drowned when those ice sheets melted and the sea returned.

The raising of islands around ancient, planetoid-impact explosion basins is a worldwide phenomenon during all ice ages, just as was the raising of mountains on the continents. Ice sheets of varying sizes caused uplift of varying extents. The great mountains of the world, the Yucatan and Florida peninsulas and the large islands of Hispaniola and Cuba show the power of large ice sheets. The works of small ice sheets are shown in low mountains, offshore islands, and atolls.

The atoll was a mystery of the sea which can now be explained as an effect of these worldwide ice sheets. The process is that, when sea level falls three miles, any land at or near present sea level will support an ice sheet. Even the warmest of modern-day tropical islands become icy mountaintops after such a sea level fall. Those islands that were covered with soft sediments developed ice sheets that acted the same as their huge continental brothers. Sediments were squeezed from the center of the island to its edges. When sea level rose and the ice sheet melted, the top of the island was left as a bowl-shaped depression with a raised rim. On some islands, these rims became hills and cliffs hundreds of feet above present sea level. When the oceans returned, these rims extended above the water and became circles of islands.

The atolls show the action of the smallest ice sheets. The largest ice sheets periodically covered most of the globe. The largest of these over continents was the North Polar Ice Cap, which comprised ice sheets reaching out of the Arctic Ocean basin to cover all of Europe north of the Pyrenees and extending eastward to cover Asia, continuing well into what is now the Pacific Ocean – and covering North America. The Eurasian ice sheets bordered on several others, and the combined actions of these ice sheets through several ice ages built the great mountains of Europe and Asia. The western edge of the Eurasian ice sheets raised Britain and Ireland out of the earlier European continental shelf. The southern limit bordered on the similarly great African ice sheets. The African ice sheets flowed north in some areas, away from the Equator, because it was downhill from the center of the ice sheet pressure. For this reason, ice sheet striations in the Sahara point to a northward movement of ice.

The pressure of the African ice sheets reached across the partially drained, frozen Mediterranean Sea to reach the Eurasian ice sheets. The pressure of these two colliding ice masses raised the Pyrenees Mountains between Spain and France, the Swiss Alps and a mountain range that would become Italy when water returned to the Mediterranean. Farther east, these great collisions elevated the Carpathian Mountains, the Dinaric Alps of Yugoslavia and the Taurus, Pontic, Greater Caucasus and Zagros mountain systems near the Black and Caspian seas. The easternmost contact of the Eurasian and African ice sheets produced the rugged plateau of Iran.

The southern border of the Eurasian ice sheets also produced the great Himalaya Mountains. India, like all the rest of the world during the great ice ages, was covered with ice sheets, which produced its Eastern and Western Ghat ranges. The magnificent Himalayas were pushed from the north, however, to a considerable extent during the time of man on Earth. This movement from the north during the last ice age forced some of the older, hardened rocks over the newer sediments of the Siwalik hills in some places.

Eastward along the southern border of the Eurasian ice sheets are intersections with other great ice sheets. These ice sheets covered all of southeastern Asia and extended into what is now the Philippine Sea to cover and elevate the islands of Indonesia, Borneo and the Philippines. The border between these ice sheets and the Eurasian ice sheets elevated the South China Highlands.

The eastern margin of the Eurasian ice sheets caused fascinating developments in what is now the Pacific Ocean. The falling sea level emptied great basins east of the Eurasian continent. The Eurasian ice sheets extruded the sediments of these basins up over the basin rims, just as did the Caribbean and Gulf of Mexico ice sheets. This piled-up material formed mountain ranges which became island chains when the oceans returned. These island chains, also known to geologists as the 'andesite line', include the Ryukyu and other islands of Japan, the Kurile Islands and the Kamchatka peninsula.

The ice sheets squeezed up mountains and highlands wherever they rested on soft material. But, of course, there were many variables which made the type of structure that was formed quite different in different places. The thickness of the ice sheet, the thickness and chemical makeup of the sediments, the underlying structure and the presence or absence of bordering ice sheets all influence whether or not mountains will be forced up.

The Swiss Alps, for instance, are an interesting variant. Instead of being formed of soft sediments elevated by pressure of magma from below, the Alps appear to have already been hard at the time that they were elevated. Under the pressure of ice, gigantic slabs of rock were uplifted and thrust over younger strata.

This same type of movement of already hardened strata can account for the formation of one type of plateau. In this case, the fluid magmas or muds, instead of breaking or folding the strata above them into mountains, simply elevate large areas of hard strata with little tilting. This action caused the elevation of formerly submerged peninsular Florida by the pressure of the Gulf of Mexico ice sheets. While the peninsula was held in its elevated position by the pressure of the ice sheets, the magmas or muds underneath the peninsula hardened. When the ice sheets melted and the seas returned, the peninsula remained as a low plateau slightly elevated above sea level.

All of the Americas were covered by giant ice sheets, just as were the other continents. Here, too, at various times, the ice raised tremendous mountains. The Aleutian Islands off Alaska were squeezed up originally as a ring of mountains around the ice-filled basin now known as the Bering Sea. The ice sheets also formed the Brooks and Alaska mountain ranges in Alaska and, farther east, the Queen Elizabeth and Parry Islands in northern Canada. To the northeast, the combined forces of the polar, Eurasian and North American ice sheets elevated the coastal mountains of Greenland. The mountains thus elevated trapped the ice of Greenland and saved it from the end of the last ice age until today.

The eastern side of the North American ice sheets elevated Newfoundland and the Labrador Peninsula in the north, and also raised the Appalachians and Nova Scotia along the eastern coast. The elevation of Florida, Cuba, the isthmus of Panama and the Yucatan peninsula, as well as the other islands of the Caribbean, has already been discussed in detail. The small ice sheets that elevated the mountains and islands of that area were aided by the North American ice sheet in elevating the mainland of Mexico with its mountains.

It was in the west, though, that the American ice sheets performed their greatest visible works. Along the west coast of the Americas runs one of the world's grandest mountain systems. This mountain chain, called the Western Cordilleras, runs all the way from lower South America to Alaska. It is made up of five main mountain ranges: the Andes, the Rockies, the Sierra Nevadas, the Cascade Range and the

Coast Range. It also includes Central America and the Mexican Plateau, and the Brooks and Alaska Ranges, made in conjunction with the polar ice sheets.

The great mountain ranges of the world were not built up by the actions of a single ice age. Although some ranges may have been greatly raised by only one ice age, some are the products of multiple elevations during many ice ages.

Metamorphosis Decoded

The piling up of materials is only half the story of the formation of mountains and islands. Those materials must harden into rock before the ice sheet melts, or they will merely slump back to their former positions. Uniformitarians allow millions of years for sediments to turn into rock, and have proposed several methods by which this can occur through great heat and pressure. But the process that forms ordinary concrete is much faster and is worth examining.

An excellent example of the ability of volcanic materials to be cemented into solid rock appears within the time of recorded history. When the remains of Herculaneum and Pompeii were discovered – cities that had been buried by an eruption – a story evolved to grip the world's imagination. Roman sentries had been buried at their posts. Family groups, in the supposed safety of subterranean vaults, had been cast in molds of volcanic mud cemented to a rocklike hardness, along with their jewels, candelabra and the food that they hoped would sustain them through the emergency. These people were not buried by boiling lava, for that would have incinerated their bodies, leaving no trace. They were buried in mud which later cemented itself into rock. Archeologists were able to pour plaster into the molds left by the decomposition of the bodies of these unfortunate Romans. After the rock was chipped away, the plaster left a perfect representation of the body in death.

In order to understand the process by which volcanic muds and dusts metamorphose into rock, let us examine the process by which man makes dusts and muds that turn into rock. The dusts are the commercial cements that are used for construction. One of these is Puzzolana, made from impure alumina silicate volcanic debris, and the other is Portland cement, made from sea-floor limestone. Cement is made from either of these raw materials by first heating in the absence of oxygen or a 'reducing' atmosphere to a temperature of 1000–1400° C, until the material begins to fuse into a glassy clinker. This clinker is then ground to a fine powder under circumstances that continue to exclude as much oxygen as possible. The resulting powder is commercial cement. When mixed with water, sand and gravel, this powder forms the rock that we call concrete.

Puzzolana and limestone are popular materials for making cement for construction because they can be made to harden a few hours. Probably any rock can be made into cement by the same process, but the hardening time for other materials may be many times longer. A cement that takes several hundred years to harden would be useless for a driveway, unless your family is good at planning ahead. But a cement that set in several hundred years would be just fine for building mountains, since the ice sheets would surely hold the cement in position for that long.

The process by which rock is turned into cement sounds complicated, but all of the steps mentioned could take place in an erupting volcano. Large-scale vulcanism could cover the Earth with dusts that would be available for becoming cement when they came into prolonged contact with water. Nature provides us with another form of concrete. In marine environments, the calcium from the skeletons of marine animals is normally deposited in oxygen-depleted environments. These marine muds, too, harden into stone when they come into contact with oxygen.

There is also a third type that nature makes for us. Some tropical soils are called laterite, a term which is derived from the Latin word for 'brick'. The Romans made roads from bricks cut from this soft soil, which hardened into stone in the air. Under the ground, these soils were protected from air and thus from oxidation. But once on the surface, they hardened into a rock that could not later be dissolved in water. Laterite soil in southeastern Asia was molded into figurines and stairways of the great temples, such as Angkor Wat. These moldings later hardened in air to a rock which has endured weathering to this day.

Examples of the use of natural cements are not limited to the ancient world. During World War II, on the island of Malta, air raid shelters were being dug into the limestone of that island. The workmen found that once they had dug through a few feet of hard limestone near the surface by drilling and blasting, the rest of the underlying limestone could be dug out with shovels. Earth-moving equipment was used to dig out huge rooms in this material. Within a few days, this material had hardened into rock.

A similar type of material, called coquina limestone for the shells that are found in great abundance in it, was used for building by the Spaniards who explored Florida. They used hand implements to cut the soft rock from the ground. After a time in the air, it could be used for building just as could any other limestone block.

Natural cements appear to have figured prominently in the history of man in many places. They have also affected the history of the Earth. It appears that many natural cements may have already hardened into mountains when the continents once again became habitable. But some may have had their setting held up by the cold of the ice sheets.

A geological formation called a peneplain is presumed to be the result of the complete wearing away of a mountain range by erosion. It is an area which resembles the base of a mountain range with the mountains themselves sliced off along a flat plane. But the problem with this theory of the peneplain is that there appear to be none in the modern world. Perhaps what occurred to form these flat surfaces was not erosion of mountains of rock, but instead erosion of mountains of soft mud. If an ice sheet were to push soft sediments up to mountainous size and then freeze them, the process of turning into rock might be slowed or halted. The mud beneath this frozen zone might still harden into rock, thus forming a nearly flat surface of solid material. But when the ice sheet melted, the action of the meltwater might simply wash away the frozen mud and leave that flat lower surface. So it is apparent that it need not take millions of years for mountains to harden in place. All of the materials of which they are made (magmas, limey muds and volcanic dusts) have the property of metamorphosing into rock in terms of hundreds, not millions, of years.

Humans Employ Mountain-building Secrets

Some of the great mysteries of prehistory can be solved if we assume that the materials that are now hard rock were once soft cement. In many places in South and Central America there are walls and buildings made of huge blocks. These constructions are thousands of years old, yet the blocks still fit together so perfectly that a knife blade cannot be inserted in the joints between blocks. No records of the construction methods have survived. Author Erich von Däniken finds the constructions so fantastic that he concludes that beings from outer space helped the ancients build them. But compelling evidence about their construction comes from ancient Machu Picchu, which is built around a spire of rock on a mountain peak. That the builders failed to remove the spire and use it for blocks strongly infers that they were limited to moving and stacking a softer rock around the spire.

Anyone who has worked with granite, limestone, sandstone (which appear in nature in both the hard and soft state) or concrete blocks (they too are soft before the cement sets) will immediately deduce the softness that these blocks had when they were stacked. This deduction may be made by seeing the way they fit together and the shapes of the blocks.

One need not visit the walls – a photograph clearly shows it – to see that each block is not only shaped to fit those below and on either side, but that these blocks in turn had been shaped to accommodate each new block as it was added.

One must decide between two methods. Either the ancients were equipped with tools far in advance of ours and the strong inclination to avoid simple and direct block shapes that would allow efficient assembly methods, or the blocks were in the process of hardening when stacked, and by the uncomplicated process of tilting or sliding each newly added block back and forth against those previously fitted into place, abrasion was caused to remove the high spots between them until a perfect 'fit' was achieved.

Throughout time, workmen left to their own devices have chosen the easiest and fastest methods. It is unlikely that they chose differently here.

Easter Island in the Pacific has also confounded scientists with its gigantic stone statues. These statues, made of solid rock and ranging up to 60 feet in height, have been found in varying stages of completion around the island. One such statue, for example, was found, nearly completed, lying on its back in a room that was carved out of the same stone of which the statues were made. The statue had not yet been cut away from the rock from which it was made, but was supported above the floor of the room by a narrow pedestal of rock.

The explorers of Thor Heyerdahl's group decided that in order to show the feasibility of making the giant idols, they would finish the job of cutting out a similar statue by using the stone implements which the ancient Easter Islanders then had available. They gave up after a few days, having barely scratched the rock.

The fact that their tools won't cut this rock today makes the carving of these giant idols by the ancient Easter Islanders an almost unbelievable feat. But compounding the mystery is the manner in which the statues were carved. If the rock was hard when the statues were made, the easiest way to carve them would be to cut them out of the rock immediately on the surface. Instead, the Easter Islanders excavated whole underground rooms around the idols and carved most of the features while the statues were still attached to the rock. This involved removing many times more rock than would making the statues on the surface.

If the mysteries of Easter Island are examined in the light of a massive ice age, however, many of the problems can be cleared up. The melting of the ice sheets at the end of the last ice age caused the seas to begin rising again. This caused a shrinking of the land upon which these people lived. Eventually, that shrinking caused a crisis, for the land was no longer large enough to support the population. After living on what is now sea floor for many generations, the people who would eventually become the inhabitants of Easter Island began to migrate. According to Easter Island legend, a war broke out between the two races that inhabited Easter Island. The island is so small that today the story is discounted, but according to sea-floor charts, what is now a small island was only the highest point of a large continent. Perhaps that war was a racial war of survival on a shrinking land.

As the ancient Easter Islanders retreated from the rising sea, they may have feared that one day the sea would engulf the whole island and kill them all. When the ice had melted enough for them to climb the mountain that was to become Easter Island, they may have found that the ice and volcanic dust had left the island with a frozen material which could be worked easily with flint tools and which would harden into rock after some exposure to warmth and air. With such a material, the excavation of a room around the statue becomes more comprehensible. It was necessary to excavate a room around each statue so that air and warmth could reach all sides of it so as to harden it for transportation. Also, the softness of the material made the excavation of such a room an easy matter.

After the giant statues had been carved and their cement had hardened, they were moved to the locations where we find them today. The giant idols with their stern expressions were placed along the seacoast to glare at the rising waters, possibly in hopes of frightening them or their enemies away.

Nearly all of the complete idols that are found today face the ocean.

This reconstruction of the history of Easter Island is pure speculation, but it is based upon fact. There is evidence that the seas were three miles lower during the ice ages than they are today, and this would make Easter Island the highest point of an area three times the size of modern South America. Rising seas would have caused a scramble for arable land, and the Easter Islanders still have a legend of a great war that exterminated a whole race. Furthermore, natural cements have been found throughout the world, protected from oxidation by burial, but which will harden into rock when they are exposed to air.

Add to this the question of how the ancients carved the giants with only stone implements, and also the question of how an island that today can only support a few thousand people could have had a large enough population in the past to permit work on idols as well as the necessary work of farming. The contrasts of an ancient and modern Easter Island lead to the conclusion that things were very different when the giant stone statues were raised.

Similar mysterious constructions in Peru may be explained if materials that are now hard rock were once soft sediments. A great city, Tiahuanacu, stands in the Andes mountains at an altitude of 12,500 feet. Agricultural terraces line the mountain above the 18,400-foot level, which marks the snow line at this latitude. No one today lives at Tiahuanacu; no record was left by its builders as to why it was located here. Much of its construction is of giant blocks of stone, which is difficult to understand at such an extreme altitude. The city stands mute, keeping the secret of its construction.

Some have suggested that the construction of the city is not at all remarkable, but that it was constructed on the shore of the Pacific by people who lived there. Later, in a giant catastrophe, it was catapulted to its present altitude as the Andes sprang from the Earth.

It is possible that this is what happened. Indeed, it is

certain that the Andes have undergone some elevation in the time of man, for there is other evidence of this elevation in the form of beach lines and fossils. But there is a flaw in this line of reasoning. If the mountains were thrust up in a great spasm of the Earth, why was the city not destroyed completely? Surely a change that violent would cause the mountains to shake off the city as a dog shakes off water. And if the area was not already mountainous, why (and how) were agricultural terraces built at all? It would seem that there was already some elevation of the area when humans began to build the city.

There is another mystery in the Peruvian Andes. Two fortresses, Ollantaytambo and Ollantayparubo, stand at high altitudes and are made of giant blocks of stone. The stones of Ollantaytambo stand 12 to 18 feet tall and come from a quarry seven miles away. The stone of Ollantayparubo is also not found in the immediate area of the fortress. These constructions stand high in some of the most rugged mountains in the world. Ollantayparubo is at an altitude of 13,000 feet. Carrying the stones to these locations today would require raising and lowering them thousands of feet, carrying them across canyons and swift-flowing rivers. Even today, with modern power equipment, transporting those giant blocks would be a difficult task. How did the ancient Peruvians do it with only simple tools and muscle power?

If these two fortresses and the city of Tiahuanacu were built as the ice sheets of the last ice age waned, the different circumstances which prevailed at that time might explain the motivations and means for these great constructions.

The humans who lived on the floors of the oceans during the decline of the last ice age were forced to lead a nomadic life. Because the rising of the oceans periodically drowned their cities, their mode of life was a constant retreat toward higher ground. Many generations had passed since their ancestors left the continents which are familiar to us. They might have had legends about dry continents and motionless seas, but their experience showed that the seas always rose, and that people always had to retreat before them.

The people who retreated in the direction of the Peruvian Andes were more fortunate than those (possibly their relatives) who fled toward Easter Island and Rarotonga. These islands were cut off early, and the constantly rising water shrunk the land, causing crowding and starvation. Those who approached the continents of North and South America had no such problems, even if they were not aware of their good fortune. Although their ancestors may have crossed the same territory the opposite direction many years earlier, the continental slope was unknown territory to them, so that a scouting party preceded them.

Perhaps the job of the advance party was more than just that of finding suitable territory for habitation. Perhaps the scouting force was a large one, whose job was to not only find new locations but prepare them for habitation. The force might have included the ancient equivalents of builders, surveyors, and engineers whose duty it was to actually build new cities and lay out new farmland, so that when the bulk of the population moved, they would be able to immediately take up residence and begin productive hunting and farming. This scouting-building force would be an invaluable asset for a society on the move.

Imagine, then, that the scouts sent word to the main body of that civilization that they had found the Andes Mountains. It was an amazing opportunity for people who had never lived in mountains. At that time, the Andes were far different from the rugged mountains that they are today. Some of the sediments that were elevated by the ice sheet were still soft and rounded. The valleys were probably filled with ice from the melting ice sheets, and everywhere were rivers and streams for a water supply.

The news must have been electrifying to the community. The gentle grade of the sea floor meant that every time sea level rose a few feet, the civilization had to relocate by miles. A city built on the steep-sided Andes would remain within reach of the ocean for many generations, as the sea rose thousands of feet, and still it would be unnecessary to move. It was a golden opportunity to stop the nomadic life for a while and settle in one location. And if the seas did not stop rising, it would be the last evidence on Earth of the existence of this society.

So, the advance engineers were instructed to build enduring cities and fortresses on the highest ground that they could find. To build at a lower level would be a waste of time, since they might have to move higher eventually anyway. The engineers built a chain of fortresses for defense, cities and agricultural terraces to permit the growing of food on the steep sides of the Andes.

The mountains themselves made the construction job easier. It appears that these mountains provided the sea-floor builders with a natural cement that was still frozen, but could easily be formed into blocks. Far from being a handicap, this made these materials much easier to handle. Had they been soft, it would have been necessary to transport them in vehicles which would hold them together as they were transported. But frozen blocks of sediment could be cut to approximately the proper size and shape and then simply dragged to their final locations, or carried on rollers.

Nature may have given these ancient builders another aid, in the form of standing ice. Ice-filled valleys would have been easy to cross on sleds, or even without them. The icy surface made a level path for sliding the block of frozen sediment from the quarries to the places that they were building.

After the frozen blocks of sediment were carved out of the mountains and skidded to the site of building, they were trimmed and worked into place just as huge soft bricks would be. Once in position, the heat of the Sun and the oxygen in the air combined to oxidize and harden these sediments into rock. In the process, though, the blocks first thawed and softened, so that settling caused them to fit together tightly. Looking at them today, we are struck by the appearance of slumping, particularly of the larger blocks.

It is possible that the great city of Tiahuanacu and its surrounding terraces were never used. Built by a bureaucracy that continued to react to the menace of a rising sea after that menace had ended, the city may have lain empty from the day of its completion to the present. The slowing of the rise of sea level may have simply made it unnecessary for the early Peruvians to retreat to the cities high in the mountains. It is also possible that the mysterious symbols on the Plains of Nazca, that to some represent runways of a spaceport, may simply be the community's chart of the boundaries of their sea-floor nations – paced off and permanently recorded so that their descendants could reclaim their rightful property 'when the oceans went down again'.

One theory of man's early history that has fallen into disrepute in recent years is the theory of diffusionism. This theory says that the cultures of the Earth all got their start in the same place. Originally there was one 'cradle of civilization' and that civilization spread from that one cradle throughout the world. Early anthropologists (before they were even called that) thought that the source of all civilization was the Garden of Eden. Later the most popular source has generally been the Fertile Crescent. Thor Heyerdahl went so far as to cross the Atlantic and Pacific in boats of early design in order to demonstrate that method of migration.

Although the theory is ridiculed by most anthropologists today, the theory of diffusionism refuses to disappear. It seems to many scientists that there are too many parallels among early civilizations, even those that are widely separated geographically, to be coincidental. Too many cultures seem to have blossomed at the same time. For these reasons, the theory of diffusionism persists.

The reconstruction of the history of the sea-floor people who would become the ancient Peruvians is, like the story of the Easter Islanders, purely speculative. Yet it is completely in consonance with the synthesis of great sea level changes and extensive ice sheets. Perhaps it also provides a clue to the persistence of the theory of diffusionism. In an ice age, people would be forced to migrate to the sea floors or perish. The sea floors would become melting pots, where humans from all over the world could contribute cultural developments to each other. At the end of the ice age, these people would take their newly formed cultures with them to the widely separated continents, and those cultures would have marked similarities. Perhaps the sea, as well as being the cradle of all life, was also the cradle of civilization that the diffusionists have sought for so long.

Hills and Caves, Deserts and Earthquakes

While the great ice sheets covered the continents of the world, they formed giant mountains all over the Earth. As they melted and retreated, they left other traces, many of which still affect us today. The meltwaters of the ice sheets, running back to the shrunken oceans, carved canyons that now lie deep under the ocean. Great rivers were also formed on the continents by these meltwaters, sometimes over the buried river beds of other ice ages. In the south-eastern United States, some of these old networks of river beds, totally unrelated to present drainage patterns, are the source of a common building material called 'brown river gravel'.

The melting ice sheets also left behind kames, drumlins and eskers. These odd names describe formations which in northern regions are considered as proof of the presence of ice sheets. Drumlins are teardrop-shaped hills, the blunt ends of which face the flow of ice and water, and kames are irregular-shaped hills of stratified glacial material. Eskers are sinuous ridges of ice-sheet deposits that show where rivers flowed under the ice sheets. Occasionally, too, a depression or 'kettle' is formed when a block of ice which had lain underground melts long after the disappearance of the ice sheets, causing the surface material to collapse.

Glaciologists have been overly conservative in their evaluation of the topographical features that describe the area and thickness of past ice sheets. For example, they have confined their identifications of drumlins – those hills of debris streamlined to their teardrop shape by ice flowing past on all sides – to the hills that are near the continental moraines or downhill from extant glaciers.

But if one takes the broad view of Earth and studies the topography of the Arctic, it is easy to see that the Arctic Ocean is the depressed center of what must have been the deepest part of the North Polar Ice Cap. And that cap, when it moved, was prevented from motion toward the Equator in the directions of Asia and North America by the resistance and elevation of these continents. So it slipped southward where it could, on either side of Greenland and Iceland. Greenland, of course, is obviously streamlined from the direction of the North Polar Ice Cap. And when the sea-floor topography is revealed, it is equally apparent that the now-submerged Reykjanes Ridge is the tail of Iceland's drumlin shape.

But these drumlins are too large to be included within uniformitarian limits of ice sheet height and area estimation. So they have been ignored, just as their miniature versions are ignored in places equatorward such as Florida. Nevertheless, they are there and must someday be acknowledged.

Underground caves, too, may be a product of the disappearing ice sheets. The formation of caves poses a problem for uniformitarians. Their quandary is that they propose the action of rain water, seeping through the rock, as the agent that made the caverns. But it is known that rainwater will begin to dissolve limestone as soon as it meets it just below the soil. Here the waters are quickly neutralized and, seeping deeper, can no longer dissolve the rock. Therefore waters seeping down from the surface into caves today are depositing, rather than dissolving, calcite, and building spectacular stalactites and stalagmites in the process.

The hypothesis that water can be the agent that first excavates caves and later begins to fill them up again with stalactites and stalagmites is patently absurd, but another explanation is now available. Probably the areas that are now caves were once strata of limey mud. After the ice raised them, only some portions of the limestone became hard, while others remained fluid. The melting of the continental ice sheets and the removal of their damming effect allowed these fluid areas to flow away. This left air-filled and interconnected caverns, later to be drained by streams and slowly filled with stalactites and stalagmites.

It is interesting to note that caverns have been found many feet under the ocean. If this hypothesis of the formation of caves is correct, it is just another indication that sea levels were once much lower than they are now.

There is a well-known mystery about that vast desert, the Sahara. Drawings of herds of cattle have been found on the rocks of that desert, and nearby human artifacts indicate that the drawings were probably made during the Neolithic (New Stone Age) period. But some of the paintings include the Egyptian god Set, and the cattle have disks set between their horns, just as did cattle during the Pharaonic times of Egypt. The evidence seems to indicate that the Sahara was well populated for a considerable time, from the Neolithic period right into the beginning of recorded history in Egypt. In addition, it appears that this area was sufficiently fertile to support herds of cattle and was once heavily forested. According to the Egyptian drawings, this fertility, too, persisted into historical times. What happened to change the conditions to make the Sahara the inhospitable place it is today?

In the Arabian desert, too, there are signs that that now-desolate area once supported a considerable population. In the southern part of that desert are ruins of cities and orchards, indicating that there was once plenty of water where there is now only desiccated sand and rock. The dry Gobi desert of Mongolia, too, shows signs of an early civilization and of widespread cultivation in an area that is now almost sterile. Other dried-out lands in the Americas – once wet, fertile and inhabited by man – attest to these world-wide dehydrations. Where did the water come from that made these areas livable, and where did it go?

All of these now-desert areas were covered by great ice sheets during the last ice age. These ice sheets, in turn, were 'dirty' – loaded with dusts and sands from volcanoes, the sea floor and other sources. As the ice sheets melted, the upper layers of dirt were exposed and became virgin, fertile soils. The ice sheets became covered with vegetation, and that vegetation was fed by the water of the melting ice. Remnants of the last great ice age, in the form of soil-and-plant-covered ice, still exist in Siberia, Alaska, Canada and similar places.

The melting ice provided a high water table which permitted extensive farming and herding. It did not fundamentally alter the climate of these areas. They were still areas of little rainfall, just as they are now, and that slight rainfall was not enough to replenish the water that the growing things used. In time, the water table was used up and lowered to the point where the crops and the grasses that fed the cattle could no longer reach it. The vegetation died out, and the soil lost the plants and roots that held it together. The light soil particles blew away, leaving only infertile sand behind in some places.

This falling water table also explains the remarkable wells that dot the Sahara and Arabian deserts. These complex underground excavations are of quite ancient construction, and are very deep. It is difficult to believe that any present-day tribe or society would purposely move into the inhospitable deserts of north Africa, build a series of deep wells and lead a nomadic life between them. Yet this is just the type of life that the Bedouins and the Tuaregs lead. It seems equally unlikely that the forebears of these nomads would wander into the desert and set up the sophisticated series of connected deep wells that is found there.

It is more likely that the ancestors of the nomads were the ancient farmers and herders whose ruined cities and destroyed orchards are still found there. The falling water table that destroyed their civilization did not do so overnight. When the water table first began to fall, they dug shallow wells to get at the remaining water. As the level of the water fell lower and lower, they simply deepened the wells that already existed. Although they could no longer lead the settled agricultural life to which they had become accustomed, they could still find water enough to continue existence just by deepening the wells to reach the constantly falling water table. In time, they developed this unrealistic watering system.

There is a popularly held idea that underground water supplies are created in a dynamic process in which the ground always purifies the water that falls on it and returns it to the underground water table where man can pump it up and reuse it. Under this theory, a reasonable amount of pumping would permit nature to replace the water which man has removed, and every water table would remain stable.

Instead, perhaps the great bulk of the underground water which we use so freely is the last remnant of the melting ice sheets. Perhaps the soil and vegetation of the surface of that dirty ice sheet settled on the rock as the ice melted and flowed down to saturate the underlying strata. In that case, the falling of the water table is not reversible especially in those areas where subsidence has taken place. Reducing or stopping pumping of an area would not only *not* cause the water table to rise, it would not stop it from falling.

There are many places in the world where formerly submerged land now stands at high altitudes. The south of England appears to have been elevated from below sea level to an altitude of about 1000 feet. Three forces, possibly in combination, may explain these elevations. One, of course, is the water age, which raised sea level hundreds of feet throughout the world. Another effect that would cause the elevation of sea bottoms is the force of ice-sheet mountain building. The soft, sea-floor sediments are squeezed up and hardened, sometimes at great altitudes. The third force that could cause elevation is ice-sheet rebound. The disappearance of a continental ice sheet permits the land that had previously been depressed under its weight to regain some or all of its former height. This rebound has certainly accounted for as much as 1000 feet of elevation in the region of the Great Lakes. Rebound is probably the combined effect of both water and still-fluid magmas that were displaced by ice-sheet weight flowing back into the areas from which they were ejected. These continuing flows give the uniformitarians their ideas about isostasy and subterranean convection currents, no doubt.

Sweden's Baltic Coast is rising at the rate of one-half centimeter per year, Baldwin Hills in Los Angeles has risen three feet in a century and the Buena Vista oil field in California's San Joaquin Valley has risen four feet in the same time. France is rising in the south and sinking in the north at the rate of two-hundredths of a foot per year. Uniformitarians have multiplied these measurements by millions of years and concluded that they are the modern-day evidence of the slow forces that build mountains. Ice sheets can explain the formation of mountains, however, and these slow, minor rebound and subsidence movements are just shadows of the monumental forces that an ice age can generate.

Earthquakes, too, may be caused by the subsidence and rebound of land following an ice age. Uniformitarians explain earthquakes in terms of drifting continents. They note that many earthquakes take place along faults – major cracks in the rocks of the Earth's surface. The San Andreas Fault, which runs near Los Angeles and San Francisco, has been carefully studied with the idea in mind that drifting continents provide the power for earthquakes. Scientists have noted movement of the two parts of the fault in opposite directions. They have observed 10 feet of motion in 100 years and, from that, uniformitarians have extrapolated that there has been 400 miles of that same motion in the last 22 million years. This is similar to clocking the speed of a marble rolling downhill into the Pacific Ocean in Los Angeles and then figuring how long it took that marble to arrive from New York. No one knows what, if anything, the marble or the San Andreas Fault was doing before we started observing it.

There is another problem with the assumption that the movement along the San Andreas Fault is due to drifting continents. The fault runs in a generally north–south direction, with the eastern (inland) side of the fault moving south or the western side moving north. But according to the theory of drifting continents, the convection cell which is providing the power for this motion is moving the West Coast out to sea. How this westward motion is transformed into two other motions, at right angles to the original motion, is not explained.

The motion of drifting continents is supposed to raise great 'tectonic features' – mountains, that is. The San Andreas Fault, despite the presumably great period of time during which the fault has been moving, has produced no striking tectonic features. The motion of the San Andreas Fault appears to have produced no features at all, other than itself.

Those who believe that the drifting of continents shows up along faults point to erratic boulders to indicate that the land along the San Andreas Fault has moved many miles. But the continental ice sheets were quite capable of moving the erratic rocks in California.

Drifting continents are supposed to account for the force that causes earthquakes. If continents don't drift, what, then, does cause nonvolcanic earthquakes? The settlement of the surface as the ice sheets melt and underground water flows away is not necessarily a peaceful process. Under these circumstances, the surface of the Earth would constantly be seeking new and more stable foundations. Most of the shifts involved would be minor, but occasionally a major shift would be necessary. These major shifts would be felt as major earthquakes to the inhabitants of an area.

Besides the motion of the Earth caused by subsidence, ice-sheet rebound would provide other motions to shake the Earth. Those areas which lay under thick ice sheets would tend to be depressed by them. The melting and disappearance of the sheets would allow those areas to regain some or all of their old altitude. The combination of the subsidence of land, due to ice-sheet melting and water runoff, and rebound, caused by the disappearance of the heavy ice sheets, could provide complicated and dangerous combinations of Earth motions. In addition, the mountains which were elevated by the ice sheets must have had some instability. They were originally held in place by the giant ice sheets, and when that support was removed, they would have some tendency to slump back to lower elevations, causing earthquakes and landslides. Other earthquakes have been caused by minor volcanic effects, although not all have resulted in the formation of volcanoes.

If nonvolcanic earthquakes were largely caused by the actions of the ice sheets, we would expect that earthquake activity would tend to decline as the last ice age recedes further into the past and the surface of the Earth tends to gain more and more stability. Indeed, this appears to be the case. According to Roman records, 57 earthquakes were felt in that city during the year 217 B.C. The ancient Chinese devoted much of their scientific effort to recording and trying to understand earthquakes. Today, although microseisms constantly jiggle the delicate equipment of the earthquake watchers, it is rare for any given locale to have frequent discernible quakes that are not directly connected with vulcanism.

The icy blasts of the last ice age exterminated the mammoths, but that was only part of the catastrophe. The continued cold killed all but the most fortunate life forms – those with access to the sea floors – and even today that last ice age occasionally claims a toll in human lives. The ice sheets left behind hills and mountains of all sizes and descriptions, and today many people live on these piles of sediments. But not all of them are as stable as they appear. Some hills contain sediments that did not harden completely. Held in place by other hardened layers, they have remained standing for thousands of years, but they are really only time bombs waiting for the proper trigger.

A layer of dried clay among hardened layers of rock provides a stable base to build on. But if that clay is soaked by continued heavy rains seeping through the strata above it, it can return to the slippery, unstable state in which it was originally deposited. A small fracture of the strata above it can cause all the layers to slide right off a hill. Indeed, once the vibration of such a landslide begins, the upper rain-soaked layer of soil may fluidize and flow downhill as a river of wet earth. This happens in many years when the spring rains soak Southern California, and has occasionally destroyed whole towns.

The drifting continents theory requires its adherents to have great faith, for many of the assumptions upon which that theory is based contravene observable fact. Rocks do not fold; they break. The Earth is not a liquid mass; it is a solid that is stronger than steel. Convection currents are an impossibility. The ice-sheet theory of mountain building does not require any faith, however, for it is based on observable facts. The evidence of the sea floor shows that sea level has many times been three miles lower than it is at present. The missing water was piled on the continents in ice sheets averaging six miles in thickness. The ice sheets raised and molded sediments and magmas into mountains, and those materials hardened just as both natural and artificial cements do. Above all, the ice-sheet mountain building hypothesis suggests that the formation of the Earth's features may be explained without violating the laws of physics.

10

The Past Reveals the Future

The Big Picture

Ice ages have probably begun when the Earth received less heat from the Sun's rays, and volcanic activity may have been the source of dusts that shaded the Earth. But what could have initiated outbreaks of vulcanism violent and extensive enough to cause ice ages? One answer is suggested by the planet Jupiter.

Approximately every 10 years, Jupiter suddenly gives out great bursts of energy. Heat, radio waves and radiation stream away from the planet in quantities much greater than usual. This cycle is very regular, and astronomers can predict when the next outpouring of vast quantities of energy from Jupiter will occur. However, it is not clear what causes these eruptions of energy. Perhaps eddy currents developed by Jupiter's motion through the electrified solar wind cause a buildup of heat under the planet's surface. In turn, this causes thermonuclear reactions which earthbound astronomers detect as blasts of radiation.

If the heat built up by eddy currents causes Jupiter to erupt every 10 years, perhaps this has also occurred on the Earth. The smaller size of our planet has extended the time between eruptions to thousands of years, but the mechanism may still be the same. Eddy currents have built up heat in the Earth's mantle until that heat and pressure which went with it caused fission-fusion reactions to create giant volcanoes. This violent volcanic activity may account for the thousands of calderas that dot the Earth.

The buildup of pressure that caused the catastrophic

vulcanism may have taken many years to relieve. The result may have been 'volcano ages', during which the sky was darkened for hundreds of years by dusts of volcanic origin. Such darkening of the sky would certainly be sufficient to begin and sustain an ice age.

Visualize a volcano many hundreds of times more powerful than any ever seen by modern humans. They have existed. The caldera left by the explosion of Krakatoa is not the largest. Even so, Krakatoa sent dusts 17 miles into the atmosphere. The ancient giants could blast dusts and gases hundreds of miles into space. What would have been the results on Earth of such a titanic volcanic jet? It would have literally blown a portion of the Earth's atmosphere out into space. The air thus ejected would not disappear, however. It would radiate its heat into space, its temperature falling to hundreds of degrees below zero. When gravity overcame its momentum and it returned to the Earth's surface, it was an icy, lethal gas that extinguished all life in vast areas. Perhaps it was this that preserved the meat of the mammoths of Siberia.

Regardless of what caused the shading of the Earth, the results of that shading are predictable. The oceans continue to evaporate at their normal rate of more than a half-inch per day, allowing the formation of rain and snow. This rate of evaporation may even have been accelerated because volcanic lava flows on the ocean floors add extra heat to the oceans.

The shading of the Earth causes the continents to cool down, which in turn allows snow to accumulate over large areas. Due to the lowered temperature, the snow does not melt. As a result, sea level falls as more and more of the water that was formerly held by the oceans is accumulated on the land. The falling sea level accelerates the onset and growth of an ice age. As the sea level falls, so does the snow line. As the snow line falls, the area of land that can hold unmelting snow fields rapidly expands. The expansion of area above the snow line causes less and less of the water that evaporates from the oceans to return to the oceans as runoff. The expanding snow fields and falling sea level reinforce each other, so that the shrinking of the oceans proceeds swiftly.

Eventually the oceans shrink to mere shallow seas in the deepest parts of their basins. Much later in those shallow seas, microscopic life forms live and die for hundreds of years, leaving their shells to form limestone floors. This limestone covers the deep-ocean ooze that was deposited when the sea levels were high. Most of the dry area of the former ocean floors becomes forests and prairies, and the shrunken seas are no barrier to migrations between all the continents, except Australia.

On the continents, vast ice sheets develop on all the former land areas, from the poles to the Equator. These ice sheets provide the necessary pressure to convert continental forests into coal, and the necessary conditions for the freezing, accumulation and burial of the bodies of untold millions of animals which later become phosphate deposits. In formerly water-filled basins, the ice sheets precipitate evaporites and dusts to seal up the cubic miles of living things that are trapped there, and to convert them to oil and gas. The great weight of the ice sheets also raises entire mountain ranges on the continents, and great island chains elsewhere.

The evidence from the sea floors indicates that sea level has, for long periods of time, been three miles lower than it is now. That fall reduced the area covered by the oceans from two-thirds to one-third of the area of the Earth. The water which had covered two-thirds of the Earth now lay as ice on that one-third that had formerly been continents and continental seas, gulfs, and shelves. The three-mile fall in sea level caused ice sheets averaging six miles thick to cover all the continents. In some places, those ice sheets were as much as ten miles thick.

The ice sheets raised mountain ranges of soft sediments and magmas to high altitudes. The altitude of the ranges was limited, however, by the height of the ice sheets that raised them. Held in place by the ice sheets, the soft magmas and sediments metamorphosed into rock. Those parts that did not completely harden before the ice sheets melted were washed away, leaving the great underground caverns and peneplaned mountains. Some half-hardened material remained in place to be used as building blocks by the ancients.

While the sea levels fluctuated about their lowest ebb, surf action eroded and deposited material to form a broad platform. This platform, now called the continental rise, surrounded the shrunken seas – just as a higher platform, now called the coastal plain and continental shelf surrounds the oceans when sea levels are high. The pounding of the surf stirred up the sediments, depositing them in stratified layers according to coarseness. These well-sorted beds are now identified as turbidites.

Ultimately the shading that caused the ice age disappeared and the heat of the Sun warmed the Earth again. This may have taken hundreds, or even thousands, of years. As the ice sheets melted, they poured enormous quantities of water back into the ocean basins. In doing so, they carved canyons, now deep under the ocean, that are greater than any seen on land today. As the ice sheets continued to melt, sea level rose to its former level. As it moved up, wave action on the edges of the then-expanding seas left other turbidites on the basin slopes.

The topography of the sea floor implies that periods of low sea level have alternated many times with periods when sea level was at its present height. The ocean-floor layers of limestone which alternate with normal deep-sea deposits prove that sea level has fluctuated widely many times. Dr Emiliani's research confirms the frequent nature of ice ages, proposing that there have been twenty in the last 700,000 years alone.

But what does this mean to man? The frequent extinctions of well-adapted species and the sudden bursting upon the scene of many new species at the same time suggest that there may be a relationship between the catastrophic ice ages that the Earth has suffered and the process of evolution. Have the ice ages also changed humans?

The Birthplace of Humanity

The job of an anthropologist is difficult, for he must attempt to reconstruct the lives of people from the scantiest of evidence. Indeed, the most valuable find that an anthropologist can make is an ancient garbage dump. There an accumulation of artifacts may be found which can be analyzed. From those artifacts can be deduced the type of society that produced them.

But this sort of scientific detective work is hampered by many things. The rarity of sites of prehistoric habitation makes the evidence inconclusive. For instance, finding a fishhook at a particular location is a good indication that the people who once lived there were familiar with fishing. If the anthropologist finds no evidence of hunting or farming, he may conclude that the society in question was primarily made up of fishermen, and construct a possible life for the society on that conclusion. But the conclusion may be entirely false. The presence of a fishhook and the absence of other implements may not indicate that the people in question had no knowledge of hunting or farming. Indeed, it may mean that hunting and farming were of such paramount importance to them that they never lost or left behind the tools of those occupations. To them, perhaps, the loss of a fishhook was of so little importance that they did not look for it and, as a result, the anthropologist was able to find it thousands of years later.

Anthropologists also attempt to show the advance of culture in a society by analyzing the artifacts of that society and drawing conclusions about the cultural sophistication that they exhibit. But such an analysis must reflect the prejudices of the anthropologist, not those of the society in question. For the anthropologist to conclude that the development of one kind of tool indicates a certain level of culture may be false if the society places more or less importance on the tool than does the anthropologist. A society that had little use for trees would make few improvements on axes, and if an anthropologist used the development of the ax as an indicator of the level of culture, he would probably estimate that the society was of a lower order than it actually was. Similarly, labeling a certain type of tool, pottery or weapon as 'primitive' or 'advanced' reflects the taste of the anthropologist, not the culture of the makers of those artifacts.

Anthropologists sometimes use the methods of treating the dead of a society as an indicator of their cultural evolution. Societies that preserve their dead or bury bodies in a consistent or ritually significant manner show evidence of a belief in gods or an afterlife. But a society that cremates its dead may have no less significant beliefs, even though no record of their dead is to be found. The absence of buried bodies does not necessarily indicate a low level of culture.

Although we find, in some places, vast deposits of fossilized remains of now-extinct animals, remains of our ancestors are extremely rare. The location of new human fossil remains is such a rarity that the finding of a single fragment of bone can cause world-wide headlines. The life of our ancient ancestors gives a clue as to why their remains are so infrequently found.

Our Stone Age ancestors stood on the brink of humanity and civilization. Theirs was not the life of a naked savage, aimlessly wandering about the Earth, randomly grubbing for scraps of food. They were organized into small bands that hunted, fished and lived collectively. They used tools for hunting and building, had the rudiments of language and cultural rules and regulations, and were well disciplined in the skills of staying alive.

This rugged outdoor life was not conducive to the preservation of their remains. In those ancient times, these ancestors of today's humans probably had the animal's regard for the dead, simply leaving them where they fell. Since they were constantly on the move, they had no central burial place where we can go to find remains. Probably most just disappeared, their flesh eaten by scavengers and their bones left to bleach and crumble in the heat of the Sun.

Scientifically known as *Homo erectus* or *Homo pekin-*

gensis, they were smaller than modern humans, perhaps five feet tall. They had coarse features, with low foreheads, heavy ridges above the eyes, and receding chins. These forerunners of modern humans also had large, protruding teeth. Although their scientific names refer to them as humans rather than apes, their brain capacities were less than three-fourths that of modern humans. It is interesting to note that *Homo erectus* was originally called *Pithecanthropus,* a term constructed from two Greek words meaning ape-man.

Despite the limited intellectual capacity of these subhumans, they were not lacking in cunning. They were canny hunters, possessed of greater brain power than any other creature on Earth up to that time. They had some creative and learning abilities, although these were limited by the size of their brains. They were superb physical specimens, due to the constant physical effort required just to survive and find food. Their stamina and resistance to the elements were remarkable. They were well acquainted with life out of doors and with the skills required to live off the land.

These subhumans were on the Earth when a great ice age struck. Many animals and subhumans froze to death almost immediately, and many species were rendered extinct. Instead of hiding and huddling in caves with other animals waiting to be slowly frozen to death, some of these subhumans struck out for new territory which their cunning and budding intellect would make livable. Those bands that lived near the seacoast and the Equator had a particularly good chance of surviving.

As the continental ice sheets grew and the oceans shrank, the bands of subhumans followed the receding water line for survival. If they had not been fishermen before, they undoubtedly acquired that skill in order to find food. Some forms of game survived the first cold blast and accompanied the subhumans in their retreat from the freezing continents, but the most reliable source of food for them was the ocean.

As the subhumans followed the shore to ever lower levels, farther and farther from the old continents, they walked over tremendously fertile soil, made of volcanic ash and decomposed sea life. Had they been farmers, no food problem would ever have arisen for them. But they were not, and as a result, they continued to have to put all their physical efforts into the acquisition of food. Sometime during this sojourn on the sea floor an amazing thing happened. The subhumans who left the continents underwent a metamorphosis and became modern people. The means by which this happened is mysterious, shrouded both by time and the fact that this change took place in areas that are now miles under the ocean. Perhaps the stresses of the catastrophe caused them to change form. Perhaps, as Erich von Däniken has suggested, visitors from another planet caused the change from subhumans to true humans. Perhaps the Biblical Genesis is correct.

In any case, the change was great. Humans became larger and stronger, and their spines changed shape so that they could hold their heads erect. The oral cavity and teeth changed, allowing more room for the movement of the tongue. This made speech easier and clearer for the new modern people. Finally, and most importantly, the braincase of the skull grew tremendously. Suddenly, people developed the intellectual capacity to create great civilizations.

The halt of the growth of the continental ice sheets and concomitant halt of the shrinking of the seas made permanent settlement possible for the new race of people that populated the sea floors. Conditions made it a much easier life than their apelike predecessors led on the now ice-covered continents. The soil which lay beneath their feet was excellent. The ocean floors were a fertile soil, and rains had leached out the ocean salt. Despite the fact that the dust still circulating in the upper atmosphere dimmed the Sun's rays somewhat, light was sufficient for photosynthesis to take place. The farming, when these new creatures with the tremendous intellect invented it, was good.

Even though the continents were frozen all the year around, temperatures were moderate on the sea bottom. The tops of the ice sheets were 12 miles above the surface upon which our sea-bottom pioneers lived. This means that the now-present jet stream winds were lacking. The sea bottom was like a greenhouse, with the Sun's rays warming the air near the surface and no strong winds to blow warm air away.

Cultivation of the newly claimed land was a simple matter, too. Ninety percent of the former sea bottom was plains and low hills. The rocks that traditionally have plagued plowmen lay far below the surface of the land, covered with what was once abyssal ooze, leaving a soil that was soft and easily worked. The surface would be comparable to the Great Plains of North America, but with fewer rocks and more fertility.

Thus, the sea-bottom dwellers had the prime ingredient for the development of a truly advanced society: abundant food. Not until people have the luxury of devoting less than 100 percent of their time to the acquisition of food can they develop the intellectual foundation for a great society. Freed from the tyranny of constant hunting for food, this society developed government and laws, religion and morals, and scientific developments which the world had never seen before.

It appears that there were at least three great centers of civilization at the sea bottom: one in the Atlantic Basin on either side of what is now the Mid-Atlantic Ridge; another in the eastern Pacific, centered perhaps in the vicinity of Easter Island; and the third in the Indian Ocean Basin off the east coast of Africa. Very likely there were others, but these are the most obvious ones. Each is a shallow area in a great ocean now, but in ancient times each was a plateau amidst fertile valleys and shallow seas fed by glacial rivers.

Thor Heyerdahl's books, *Kon-Tiki, Aku-Aku* and *The Ra Expeditions,* attempt to account for remarkable similarities among societies whose lands are separated by vast oceans. Heyerdahl demonstrated that it is possible to cross these great distances using ancient vessels. To him, this proves that ancient sailors visited both sides of the Atlantic, thus accounting for many societal and physical similarities between ancient Central America and ancient Egypt. Similar voyages to Easter Island from South America and to the Tuamotu Archipelago demonstrated that there was once communication among these areas.

Heyerdahl's curiosity about the similarities of societies now remote from each other is well-founded, but there is a simpler explanation than flotillas of colonists crossing thousands of miles of ocean with no navigational aids. In fact, the communication was on dry land, across vast former sea bottoms.

As the dust cleared from the Pleistocene skies, the Earth began to warm again and the continental ice sheets began to melt. This meant that the areas of the sea bottom that had nurtured the human race and civilization began to be covered by steadily rising seas. As people retreated from the rising waters, they moved in a dispersive pattern. In the Atlantic Basin, some headed east to Europe, some southeast to Africa and others west to North, Central and South America. Some may have retreated to the plateau that is now the Mid-Atlantic Ridge and attempted to survive in the area of the Azores. Similar dispersals of people occurred from the other 'cradles of civilization' in the Pacific and Indian Ocean basins.

Anthropologists tell us that Cro-Magnon people, the first *Homo sapiens*, appeared in southern Europe about 35,000 years ago. These are undoubtedly the immigrants from the sea floor. The figure of 35,000 years is based on radiocarbon dating – and lies near the limit of that system's reliability. The period may be considerably shorter. Bishop Ussher calculated the Creation to have started at 9 A.M. on October 26, 4004 B.C. The Mayans began their calendar with a date equivalent to August 11, 3114 B.C., although they were able to leave cryptic glyphs naming dates millions of years before that. Plato gave the time of inundation of Atlantis as 11,500 years before the present. Professor Cesare Emiliani finds that a great sea-floor temperature change occurred 11,500 years ago. Most great river deltas are too small in area to be much older than 15,000 years. Niagara Falls started to march backward to Lake Erie about 5000 years ago.

The immigrants built cities which have been totally destroyed. Some portion of this destruction is the natural result of the erosion of nature, but another factor must be taken into account, too. Throughout history, individuals have attempted to destroy those parts of the past that they considered derogatory or uncomplimentary. Conquerors routinely steal or destroy the great works of vanquished civilization. Monarchs destroy or deface monuments of previous kings in order that their own glory does not suffer by comparison.

Under these circumstances, most of the first continental cities of these sea-bottom dwellers appear to have been irretrievably lost. But traces of their existence can be found in our cultures, and someday physical evidence will be found. Just as did the Stone Age humans who retreated from the snow of the continents, the now-civilized societies which retreated from the rising seas brought their culture with them. They returned to the now-hospitable continents with languages, tools, architecture and religions which had been developed on the sea floor. Although these were to be modified as time passed with these groups separated from each other, remarkable similarities would persist, some right up to the present day.

The Answers

Questions were raised at the beginning of the book, and it was stated that the answers would show that the hypothesis of drifting continents should give way to the synthesis that vast ice sheets and great sea level changes better explain the physical evolution of Earth's surface. Here is a tabulation comparing the two different views.

DRIFTING CONTINENTS	ICE-SHEET-TECTONICS SYNTHESIS
PLANETARY HEAT AND MAGNETIC FIELDS?	
Original heat of formation plus radioactivity from unknown source. Magnetics caused by mysterious fluid rock circulation.	Observable planetary momentum decay is being electrically converted into crustal electromagnetic fields. Radioactivity may be from exploded planet Aster. Both yield heat.
EARTH – COLD OR HOT CORE?	
All molten – with convection currents moving continents, etc.	Seismologists have proved that no convection currents exist – that all vulcanism and internal heat evidence could be in the top 17 miles of crust.
EARTH'S CIRCULAR FEATURES?	
No explanation.	Comparison with Mercury, Venus, Moon, Mars – and probably all when visible – shows consistent planetoid-impact explosion basins same as Earth's. Interlocking circles across continents and ocean floors prove continents have not drifted.

TWO PLATFORMS – ONE NEAR PRESENT SEA LEVEL, OTHER THREE MILES BELOW?

No explanation.

The platforms are wave-cut terraces, proving that sea levels lingered at these altitudes for long periods.

PLACER ORE DEPOSITS?

No explanation.

Catastrophic vulcanism smelts the ores and jets them into the atmosphere. Air sorting and surface melting of ice sheets wash them into lodes or placers.

MOUNTAIN RANGE ROCK FOLDING?

All rock flows like Silly Putty if kept under pressure for millions of years.

Cold, crystalline rock may slip on crystal planes in terms of inches, but the *miles* of flow in mountains shows that material in 'mud' form like wet concrete metamorphosed (hardened) after being folded.

MOUNTAIN BUILDING?

Drifting continents squeeze up mountains when they run into each other.

Gravity and the weight of thick ice sheets force muds and magmas aside and upward.

DRIFTING CONTINENTS	ICE-SHEET-TECTONICS SYNTHESIS

EVAPORITE BASIN DEPOSITS?

As continents advance and retreat from each other, they raise and lower the floors of seas and gulfs between them.	The dozens of cycles of basin floor deposits match the dozens of cycles of ice-sheet lowering of sea levels. It is the ice-sheet action that precipitates the deposits.

ORIGIN OF OIL AND PHOSPHATES?

Ignored.	Ice-sheet extinction of marine life forms oil, and of land animals forms phosphate.

COAL ORIGIN?

Ignored.	Ice-sheet burial and compression of forests form coal.

NATURAL 'REDI-MIX' CONCRETES?

Ignored.	Frozen or de-oxidized soft sediments may have been available to early humans to use as self-hardening bricks.

SUBMARINE CANYONS AND VALLEYS?

Canyons result from mud slides. Valleys are hard to explain.

Submarine valleys and their tributary canyons show the routes that ice-sheet meltwaters carved.

ALTERNATING LIMESTONE AND ABYSSAL OOZE LAYERS THREE MILES DOWN ON OCEAN FLOORS?

Each 'plate' drifted back and forth under the Equator many times.

Alternating shallow-water evidence (limestone) and deepwater evidence (ooze) clearly show the great changes in ocean depth – and frequency – caused by ice sheets.

MAXIMUM HEIGHT OF NON-VOLCANIC MOUNTAINS?

Ignored.

Ice sheets cannot form above the atmospheric climate zone (13 miles) so that mountains cannot be elevated above ice-sheet-plus-rock-counterweights that exceeds this height.

DRIFTING CONTINENTS	ICE-SHEET-TECTONICS SYNTHESIS

EXTINCTION OF OLD AND SUDDEN APPEARANCE OF NEW SPECIES?

Millions of years of sedimentary rock layers must be regularly absent.	Old species were quickly extinguished by ice sheets. New species proliferated on dry sea floors and 'instantly' flooded the landscape, as the fossil record clearly show.

INTERCONTINENTAL MIGRATIONS AND AUSTRALIA'S EXCEPTION?

The continents were one that broke up, except Australia which wandered a bit before joining Indonesia.	Topography shows that while low sea levels permit migrations between most continents, deep waters isolate Australia.

A decision between the hypothesis of plate tectonics (drifting continents) and the ice-sheet-tectonics synthesis is difficult to make when the debate revolves about such difficult-to-prove items as a hot or cold Earth core; the reasons for circular features and high and low platforms on Earth's surface; whether mountain ranges have been raised by the sea floor sliding under them or by continental muds flowing out from under ice sheets; submarine canyon genesis; life form distribution; and the significance of sea-floor magnetic parallelism.

But an incontrovertible item of evidence for the synthesis is in the sheets of coal which lie just under the surface of millions of square miles of the interiors of the continents.

The word 'plate' in plate tectonics represents the rigid 'shield', 'buckler' or 'craton' which is the interior of each continent and that cannot, according to the hypothesis, be distorted by convection currents. Therefore drifters propose that around these unyielding disks the fancied convection currents must raise mountains, but that no mountain building forces can penetrate the interiors of these plates.

Most bituminous coalfields lie close enough to the surface so that ordinary earthmoving equipment can remove the soil and rock covering them in the process called strip mining. The layers of rock above and below the coal can be proved to have been laid down – as was the coal – in great flat sheets that have not subsequently been distorted by any horizontal pressure. But it can also be proved in the laboratory, beyond any reasonable question, that pressure equivalent to a layer of granite at least 7500 feet thick had to be laid over these millions of square miles of peat before the conversion into bituminous coal could take place. Further, it has been proved that each layer of coal was converted by such pressure, that the pressure was then removed, that a new layer of wood and plants flourished over the coal – and the scene was reset for pressurization.

No alternative hypothesis for the gentle onset and disappearance of the source of the pressure will stand up to examination. If one entertains the idea that more than 7500 feet of sand or mud regularly buried continental interiors and then was washed clearly away, leaving not one grain as evidence of its passage, one is faced with the dilemma – from where did these masses come, and to where have they gone?

It is certain that only one pressure source meets all of the criteria. Ice sheets three times 7500 feet thick can arrive as snowflakes, silently burying and freezing the forests, bringing with the snow the dusts that encapsulate the resulting peat, and growing gradually in thickness until the mass of ice causes the peat to crystallize and condense into bituminous coal.

Only ice, standing without perceptible horizontal movement in the basin-like 'plates' of the continents, can subsequently melt and run away leaving no trace of its presence – except the erosion pattern of the continental margins, the submarine canyons and valleys.

The choice between the hypothesis and the synthesis is therefore easily made. It is certain that the force of previously unsuspected volumes of ice pressure has been placed on the continental basins many times. It is not at all certain that convection currents exist in Earth's core, or that they could move the sea floors and continents if they did. Since either pressure source provides the force to raise mountains and disperse populations, the one that can be proved to exist must be preferred to that which is purely conjectural.

Destination Unknown

One of the many thousands-of-years-old stories with which nearly all Westerners are acquainted is the Biblical story of the Great Flood. Many stories in the Bible are miraculous, but this one may have its root in simple and scientific fact. The rising of the oceans and the destruction of the homeland of the ancient Hebrews would have made a powerful impression. It is not surprising that this would find its way into their Bible in slightly altered form.

Nor is it surprising that stories of a great flood are found in the folklore of societies all over the world. The Babylonian epic *Gilgamesh* includes the story of a great flood, as does Greek mythology. Across the Atlantic, the Mayans, Aztecs and Amerindians all have tales of a flood in their past. The southeastern Asians and the Chinese include a flood of monumental proportions in their histories.

Were these flood stories a retelling of the inundation of land from the rising seas caused by melting ice sheets? It seems probable, because of their widespread occurrence. Societies throughout the world were shaped, and indeed created, with the rising and falling of the sea and the freezing and thawing of the continents. Today we live thousands of miles away from, and miles above, the cities of our ancestors. Those cities are beneath the deepest parts of the oceans. Someday we will be able to explore the sites of these civilizations that set the moral and philosophical tone of all the nations of today's world.

On the broadest scale, we can now see that all of Earth's life forms are so overstressed by the cycles of the ice ages that each time they are substantially annihilated, mutated and forced to migrate far from lands they knew and enjoyed.

The return to a water age, although slower, may be only a little less catastrophic in that again all life forms are driven from their latest Gardens of Eden and must compete and wrest food from much-diminished land areas.

Too harsh a reality? As long as we must see innocents suffer and die, we must accept that our gods act with dispassionate deliberation. We must look upon this solar system in the same way.

If we set out to conceive a space ship that must transfer life forms from one part of the universe to another – and had the capability to use every resource that the laws of physics allow – what better vehicle could we form than one which allows its inhabitants to ride on the outside, where the whole of space may be contemplated?

How would we better ensure that life-sustaining light and warmth would be provided for the eons of time the journey would take? How else would we periodically cleanse the surface of its old debris and re-cover it with life-sustaining nutrient soils?

How would we assure that the life forms would be refined and improved if only the largest and best adapted were to prosper? From time to time it would be necessary – for the benefit of the ultimate objective – to wipe the slate nearly clean, leaving only a few for seed, and providing for their mutation to become more proficient. And would we be sufficiently advanced to provide things of beauty that have no function at all? While the end of such a voyage cannot – and should not – be predicted, it is useful and necessary to observe that this space ship Earth may have been put on 'automatic' controls. And that if it was given the dispassionate instruction only to deliver 'life forms', it may not appreciate what truly marvelous fellows we are.

Perhaps the Biblical admonition '. . . fill the Earth . . . and *subdue* it . . .' is advice that this free ride can terminate for us at the next slate-wiping, unless we are highly enough developed life forms to learn how the system operates – and control it so that we reach that final destination.

Index